Construction technology

Volume I

This volume should be read in conjunction with the requirements of
The Building Regulations 1985.

Construction technology

Volume I
Second Edition

R. Chudley M.C.I.O.B.

Chartered Builder

Illustrated by the author

Longman Scientific & Technical,
Longman Group UK Limited,
Longman House, Burnt Mill, Harlow,
Essex CM20 2JE, England
and Associated Companies throughout the world

First published 1973
Second edition 1987
Second impression 1988
Third impression 1989

British Library Cataloguing in Publication Data
Chudley, R.
    Construction technology.—— 2nd ed.
    Vol. 1
    1. Building
    I. Title
    690 TH 145

ISBN 0-582-42036-9

Set in Press Roman 10 on 12 point type

Produced by Longman Singapore Publishers (Pte) Ltd
Printed in Singapore.

To
MY WIFE
for her encouragement,
patience and understanding

Contents

List of illustrations

Preface

              In writing this book it is not my intention to present a comprehensive reference book on elementary building technology, since there are many excellent textbooks of this nature already in existence, which I urge all students to study.

My purpose, therefore, has been to prepare in concise note form, with ample illustrations, the basic knowledge the student should acquire in the first year of any building technology course of study.

To keep the book within a reasonable cost limit I have deliberately refrained from describing in depth what has been detailed in the drawings.

Building technology is an extensive but not necessarily exact subject. There are many ways of obtaining a satisfactory construction in building, but whichever method is used they are all based upon the same basic principles and it is these which are learnt in building technology.

The object of any building technology course is to give a good theoretical background to what is essentially a practical subject. With the knowledge acquired from such a course, coupled with observations of works in progress and any practical experience gained, a problem should be able to be tackled with confidence by the time the course of study has reached an advanced level.

Another aspect of a course of this nature is to give sufficient basic knowledge, over the whole field of building activities, to enable the technologist to hold and understand discussions with other related specialists.

Acknowledgements

We are grateful to the following for permission to reproduce copyright material:

British Standards Institution for reference to British Standards Codes of Practice; Building Research Station for extracts from *Building Research Station Digests*; Her Majesty's Stationery Office for extracts from Acts, Regulations and Statutory Instruments.

Introduction

There are, in general, two aspects of building technology:

1. Conventional or traditional methods.
2. Modern or industrialised methods.

The first is covered by the syllabus of the first two years of most building technology courses, whereas the second is an extension of the first and is covered by advanced building technology courses. This does not mean that there is no overlapping or reference to modern techniques in elementary courses. The first two years of a building technology course concentrates on the smaller type of structure such as a domestic dwelling of one or two storeys built by traditional methods. Generally it is cheaper to construct this type of building by these methods unless large numbers of similar types are being erected at the same time.

By industrial methods we mean those which are mainly composed of factory produced components to a module or standard increment such as 300 mm.

It is a fact that over half of the building contractors in this country employ only a few operatives and are therefore small firms. These firms are mainly engaged on maintenance work, extensions and one-off jobs, in the main using traditional methods and materials. It is essential that all students of building have a good knowledge of these methods and materials. For those that make their careers in the industrialised building side of the industry they will find that the majority of these systems have been developed from traditional building techniques.

The building team

Building is essentially a team effort in which each member has an important role to play. Figure 1 shows the organisation structure of a typical team and the role of each member is defined below:

Building owner: the client; the person who commissions the work and directly or indirectly employs everybody.

Architect: engaged by the building owner as his agent to design, advise and ensure that the project is kept within cost and complies with the design.

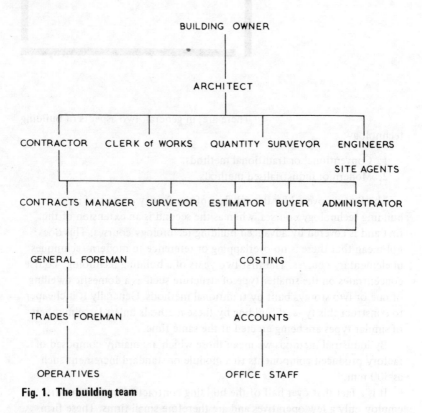

Fig. 1. The building team

Clerk of works: employed on large contracts as the architect's on-site representative. He has only liaison powers and cannot issue instructions on his own behalf; he can only offer advice.

Quantity surveyor: engaged to prepare bills of quantities, check tenders, prepare interim valuations and advise the architect on the cost of variations.

Engineers: specialists such as a structural engineer employed to work with the architect on that particular aspect of the design.

Site agent: on large contracts the engineer's on-site representative.

Contractor: employed by the building owner, on the architect's advice, to carry out the constructional works. He takes his instructions from the architect.

Surveyor: employed by the contractor to check and assist the quantity surveyor in the preparation of interim valuations and final accounts. He may also measure work done for bonus and sub-contractor payments.

Estimator: prepares unit rates for the pricing of tenders and carries out pre-tender investigations into the cost aspects of the proposed contract.

Buyer: orders materials, obtains quotations for the supply of materials and services.

Accountant: prepares and submits accounts to clients and makes payments to suppliers and sub-contractors. He may also have a costing department which would allocate the labour and material costs to each contract to assist with the preparation of future tenders and with the preparation of accounts.

Administrator: organises the general clerical duties of the contractor's office for the payment of wages, insurances and all necessary correspondence.

Contracts manager: liaises between the office and the site and has overall responsibility for the site operations.

General foreman (sometimes called the agent): the contractor's on-site representative, responsible for the day-to-day running of the site.

Trades foreman: in charge of a trade gang.

Operatives: the main work force on-site, including tradesmen, apprentices and labourers.

The size of the building firm or the size of the contract will determine the composition of any team. For the medium sized contract some of the above jobs may be combined: the surveyor may also fulfil the function of estimator.

Statutory Instruments

These are rules which are made under an Act of Parliament and are therefore legally binding on the builder and the architect. The main ones with which the builder is concerned are:

The Construction Regulations: these are for the health, safety and welfare of operatives and deal with such things as excavations, scaffolds, site accommodation and lifting equipment.

The Building Regulations 1985: these are designed to set minimum standards for buildings in the context of functional requirements to achieve acceptable health, safety and energy conservation objectives. They are made under the consolidating statute The Building Act 1984 and apply throughout England and Wales.

The Building Regulations are supported by a series of Approved Documents (AD's) which are not mandatory but give practical guidance on how to comply with the requirements of the regulations. There is also a set of mandatory rules on means of escape in case of fire.

Generally the control of the Building Regulations is vested in the local authority but a developer can opt for private certification whereby the developer and an approved inspector jointly serve an initial notice on the local authority describing the proposed works which the local authority can reject within ten days. In this option the responsibility for inspecting plans, work, site supervision and certification of satisfactory completion rests with the approved inspector as set out in The Building (Approved Inspectors) Regulations 1985.

British Standards

These are issued by the British Standards Institution, British Standards House, 2 Park Street, London W1Y 4AA. The standards are presented in two forms:

Codes of practice: these are codes of good practice in particular fields of activity such as drainage and structural steelwork.

Standard specifications: these are specifications which deal with materials and components such as bricks and windows.

Draft for development: these are issued instead of a Code of Practice or Standard Specification where there is insufficient data or information to make a firm or positive recommendation.

Published document: these are publications which cannot be placed into any of the above categories.

Codes of practice and standard specifications are compiled by specialist committees of interested parties in the particular subject of the code or specification. It must be remembered that these codes and specifications are only recommendations. Generally British Standard Specifications and British Standard Codes of Practice will satisfy the requirements of the building regulations. Copies of the codes and specifications can be obtained from the British Standards Institution, Sales Branch, 101 Pentonville Road, London N1 9ND.

Metrication

The building industry has changed to the metric system of measurement and during any period of change there is bound to be a great deal of modification and rationalisation as the designers and manufacturers adapt themselves to the new units, particularly with the recommendation for coordinated planning based on the preferred dimension of 300 mm.

Although the metric system has been adopted it should be remembered that a large proportion of the nation's building stock was designed and constructed using imperial units of measurement which are not always compatible with their metric equivalents. It can be assumed therefore that many building materials and components will be continued to be produced to imperial sizes for the purpose of maintenance, replacement and/or refurbishment.

Part I
Substructure

Part I

Substructure

1
Site works and setting out

When a builder is given possession of a building site he will have been provided with a site lay-out plan and the drawings necessary for him to erect the building. Under most forms of building contract it is the builder's responsibility to see that the setting out is accurate.

The site having been taken over, the task of preparing for and setting out the building can be commenced. These operations can be grouped under three headings:

1. Clearing the site.
2. Setting out the building.
3. Establishing a datum level.

Clearing the site

This may involve the demolition of existing buildings, the grubbing out of bushes and trees and the removal of soil to reduce levels. Demolition is a skilled occupation and should only be tackled by a skilled demolition contractor. The removal of trees can be carried out by manual or mechanical means. The removal of large trees should be left to the expert.

Building Regulation C1 'The ground to be covered by the building shall be reasonably free from vegetable matter.' This is in effect to sterilise the ground since the top 300 mm or so will contain plant life and decaying vegetation. This means that the top soil is easily compressed and would be unsuitable for foundations. Top soil is valuable as a top dressing for gardens and may be disposed of in this manner. The method chosen for carrying out the site clearance work will be determined by overall economics.

Setting out the site

The first task is to establish a base line from which the whole of the building can be set out. The position of this line must be clearly marked on-site so that it can be re-established at any time. For on-site measuring a steel tape should be used (30 metres would be a suitable length). Linen and plastic coated tapes are also available. The disadvantage with linen tapes is that they are liable to stretch.

After the base line has been set out, marked and checked the main lines of the building can be set out, each corner being marked with a stout peg. A check should now be made of the setting-out lines for right-angles and correct lengths. There are several methods of checking if a right-angle has been established and in fact the setting out would have been carried out by one of these methods. A check must still be made and it is advisable to check by a different method to that used for the setting out. The setting-out procedure and the methods of checking the right-angles are illustrated in Fig. I.1.

After the setting out of the main building lines has been completed and checked, profile boards are set up as shown in Fig. I.2. These are set up clear of the foundation trench positions to locate the trench, foundations and walls. Profile boards are required at all trench and wall intersections.

Establishing a datum level

It is important that all levels in a building are taken from a fixed point called a 'datum'. This point should now be established, wherever possible this should be related to an ordnance bench-mark. This is an arrow with a horizontal mark above the arrow, the centre line of the horizontal being the actual level indicated on an ordnance survey map. Bench-marks are found cut or let into the sides of walls and buildings. Where there are no bench-marks on or near the site a suitable permanent datum must be established. A site datum or temporary bench-mark could be a post set in concrete or a concrete plinth set up on site.

SLOPING SITES

Very few sites are level and therefore before any building work can be commenced the area covered by the building must be levelled. In building

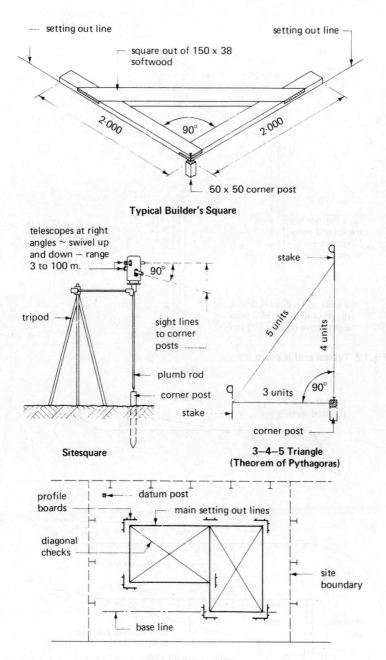

Fig. I.1 Setting out and checking methods

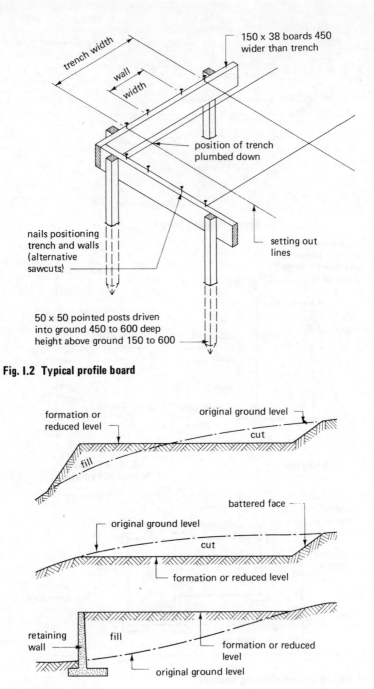

150 x 38 boards 450 wider than trench

trench width

wall width

position of trench plumbed down

setting out lines

nails positioning trench and walls (alternative sawcuts)

50 x 50 pointed posts driven into ground 450 to 600 deep height above ground 150 to 600

Fig. I.2 Typical profile board

formation or reduced level

original ground level

cut

fill

battered face

original ground level

cut

formation or reduced level

retaining wall

fill

formation or reduced level

original ground level

Fig. I.3 Sloping sites

terms this operation is called reducing levels. Three methods can be used and it is the most economical which is usually employed.

1. **Cut and fill:** the usual method because, if properly carried out, the amount of cut will equal the amount of fill.
2. **Cut:** this method has the advantage of giving undisturbed soil over the whole of the site but has the disadvantage of the cost of removing the spoil from the site.
3. **Fill:** a method not to be recommended because, if the building is sited on the filled area, either deep foundations would be needed or the risk of settlement at a later stage would have to be accepted.

The principles of the above methods are shown in Fig. I.3.

2
Excavations and timbering

Before a foundation can be laid it is necessary
to excavate a trench of the required depth and width. On small contracts
this is still carried out by hand but on large works it may be economic to
use some form of mechanical trench digger. The general procedure for
the excavation of foundation trenches is illustrated in Fig. I.4.

Timbering

This is a term used to cover temporary
supports to the sides of excavations and is sometimes called planking and
strutting. The sides of some excavations will need support to:

1. Protect the operatives while working in the excavation.
2. Keep the excavation open by acting as a retaining wall to the sides of
   the trench.

The type and amount of timbering required will depend upon the
depth and nature of the subsoil. Over a short period many soils may not
require any timbering but weather conditions, depth, type of soil and
duration of the operations must all be taken into account and each
excavation must be assessed separately.

Suitable timbers for this work are:

Scots pine
Baltic redwood
Baltic whitewood
Douglas fir

8

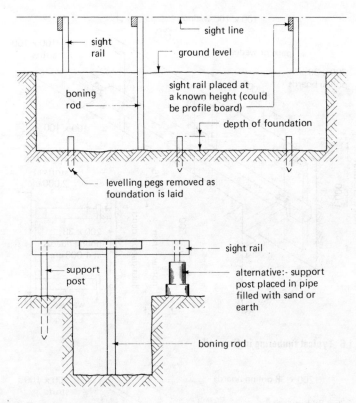

Fig. I.4 Trench excavations

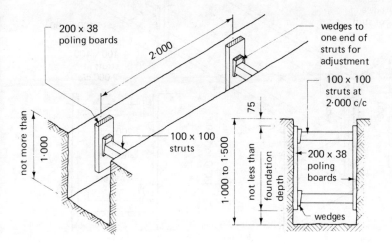

Fig. I.5 Typical timbering in hard soils

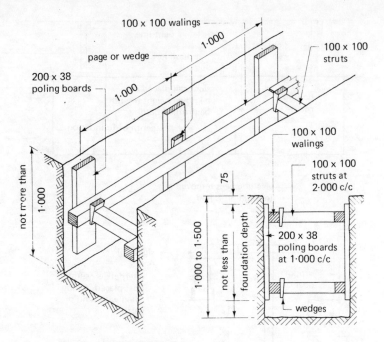

Fig. I.6 Typical timbering in firm soils

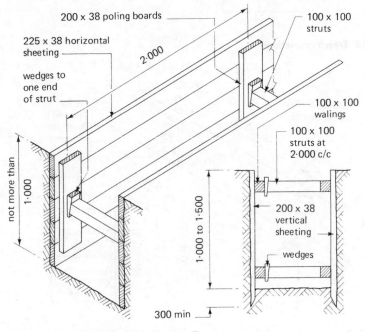

Fig. I.7 Typical timbering in dry loose soils

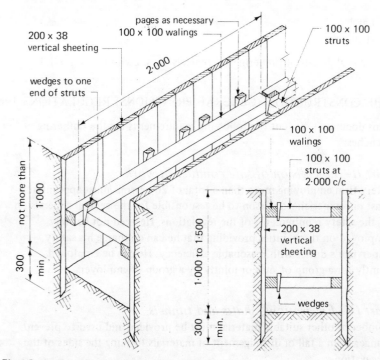

Fig. I.8 Typical timbering in loose wet soils

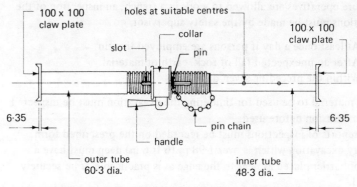

SIZE No.	MINIMUM LENGTH	MAXIMUM LENGTH
0	0·30 m	0·45 m
1	0·45 m	0·68 m
2	0·68 m	1·06 m
3	1·06 m	1·67 m

Fig. I.9 Adjustable metal struts—BS 4074

11

Larch
Hemlock.

Typical details of timbering to trenches are shown in Figs. I.5–I.9.

THE CONSTRUCTION (GENERAL PROVISIONS) REGULATIONS 1961

This document sets out the following requirements for the timbering of
trenches:

Part II–Supervision of safe conduct of work
Every firm employing more than a total of 20 persons must appoint at
least one experienced person to be responsible for the general supervision
of the safety requirements of the regulations. He need not be fully
employed on these duties providing that he can discharge his safety
supervisor's duties with reasonable efficiency. He can be employed
jointly for a group of sites or jointly by a group of employers.

Part IV–Excavations, shafts and tunnels:
Timber or other suitable material must be provided and used to prevent
danger from a fall or dislodgement of materials forming the sides of the
excavation.

Timbering must only be erected by a competent person or under his
direction.

Before operatives are allowed to work in a trench, an inspection of the
excavations must be made by the safety supervisor:

1. At least once a day if persons are employed therein.
2. After an unexpected fall of rock, earth or material.
3. Within the preceding seven days.

The material to be used for timbering the excavation must be inspected
on each occasion before used.

All reports of inspections must be recorded on the prescribed form.

Every excavation which is over 1·98 m (6 ft 6 in) deep must have a
fence or barrier placed as near to the edge as is practicable or be securely
covered.

BUILDING REGULATIONS 1985

Notice of commencement and completion of certain stages of work
Building Regulation 14 requires the local authority to be notified by
a person carrying out building work prior to commencement and at

12

certain stages during the building operations. The notice should be given in writing or by such other means as may be agreed with the local authority.

Notices are required under this regulation as follows:

48 hours prior to commencement of work.
24 hours before excavations are covered up.
24 hours before any concrete or material laid over a site is covered up.
24 hours before any drains or sewers are covered up.
7 days after completion of laying drains or sewers including bedding and backfilling.
7 days after completion of building work and/or 7 days before occupation.

In the calculation of a period of twenty-four hours or forty-eight hours no account shall be taken of a Saturday, Sunday, Christmas Day, Good Friday, Bank Holiday or a day appointed for public thanksgiving or mourning.

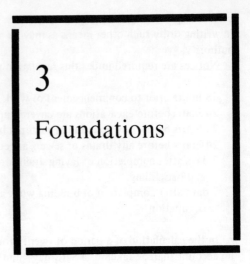

3
Foundations

A foundation is the base on which a building rests and its purpose is to safely transfer the load of a building to a suitable subsoil.

Building Regulation 4 requires that all foundations of buildings shall:

1. Safely sustain and transmit to the ground the combined dead and imposed loads so as not to cause any settlement or other movement in any part of the building or of any adjoining building or works.

TABLE 3.1 *Typical subsoil bearing capacities*

Type	Bearing capacity (kN/m^2)
Rocks, granites and chalks	10 000-600
Non-cohesive soils Compact sands Loose uniform sands	600-100
Cohesive soils Hard clays Soft clays and silts	600-0
Peats and made ground	To be determined by investigation

2. Be of such a depth, or be so constructed, as to avoid damage by swelling, shrinkage or freezing of the subsoil.
3. Approved Document 7 recommends foundations to be capable of resisting attack by deleterious material, such as sulphates, in the subsoil.

Subsoils are the soils below the top-soil; the topsoil being about 300 mm deep. Typical bearing capacities of subsoils are given in Table 3.1.

Terminology

Backfill: materials excavated from site and if suitable used to fill in around the walls and foundations.

Bearing capacity: safe load per unit area which the ground can carry.

Bearing pressure: the pressure produced on the ground by the loads.

Made ground: refuse, excavated rock or soil deposited for the purpose of filling in a depression or for raising the site above its natural level.

Settlement: ground movement which may be caused by:

(*a*) Deformation of the soil due to imposed loads.
(*b*) Volume changes of the soil as a result of seasonal conditions.
(*c*) Mass movement of the ground in unstable areas.

Choice of foundation type

The choice and design of foundations for domestic and small types of buildings depends mainly on two factors:

1. The total loads of the building.
2. The nature and bearing capacity of the subsoil.

The total loads of a building are taken per metre run and calculated for the worst case. The data required is:

(*a*) Roof load on the wall—1 m wide strip from ridge to eaves.
(*b*) Floor load on the wall—1 m wide strip from centre of the floor to the wall.
(*c*) Wall load on the foundations—1 m wide strip of wall from top to foundation.
(*d*) Total load on the foundations—summation of (*a*), (*b*) and (*c*).

The average loading for a two-storey domestic dwelling of traditional construction is 30–50 kN/m.

The nature and bearing capacity of the subsoil can be determined by:

1. Trial holes and subsequent investigation.

2. Bore holes and core analysis.
3. Local knowledge.

Approved Document A gives a table of subsoil types, field tests and suitable minimum widths for strip foundations.

Clay is the most difficult of all subsoils with which to deal. Down to a depth of about 1 m clays are subject to seasonal movement which occurs when the clay dries and shrinks in the summer and conversely swells in the winter with heavier rainfall. This movement occurs whenever a clay soil is exposed to the atmosphere and special foundations may be necessary.

Subsoils which readily absorb and hold water are subject, in cold weather, to frost heave. This is a swelling of the subsoil due to the expansion of freezing water held in the soil and like the movement of clay soils it is unlikely to be even and special foundations may be needed to overcome the problem.

TYPES OF FOUNDATIONS

Having ascertained the nature and bearing capacity of the subsoil the width of the foundation can be determined by either:

1. The minimum as given in the table to Approved Document A.

2. $\dfrac{\text{Total load of building per metre}}{\text{bearing capacity of subsoil}} = \text{minimum width.}$

Foundations are usually made of either mass or reinforced concrete and can be considered under two headings:

Shallow foundations: those which transfer the loads to subsoil at a point near to the ground floor of the building such as strips and rafts.

Deep foundations. Those which transfer the loads to a subsoil some distance below the ground floor of the building such as a pile.

Raft foundations are often used on poor soils for lightly loaded buildings and are considered capable of accommodating small settlements of the soil. In poor soils the upper crust of soil (450-600 mm) is often stiffer than the lower subsoil and to build a light raft on this crust is usually better than penetrating it with a strip foundation.

Typical details of the types of foundations suitable for domestic and similar buildings are shown in Figs. I.10-I.12.

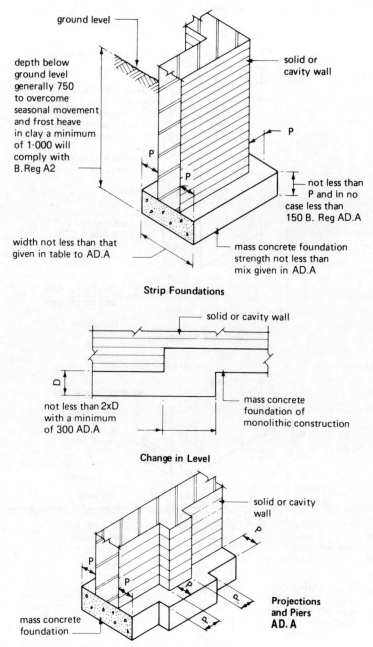

ground level

depth below ground level generally 750 to overcome seasonal movement and frost heave in clay a minimum of 1·000 will comply with B.Reg A2

solid or cavity wall

P

P

P

not less than P and in no case less than 150 B. Reg AD.A

width not less than that given in table to AD.A

mass concrete foundation strength not less than mix given in AD.A

Strip Foundations

solid or cavity wall

D

not less than 2xD with a minimum of 300 AD.A

mass concrete foundation of monolithic construction

Change in Level

solid or cavity wall

P

P

P

P

P

P

mass concrete foundation

Projections and Piers
AD. A

Fig. I.10 Strip foundations and Building Regulations

17

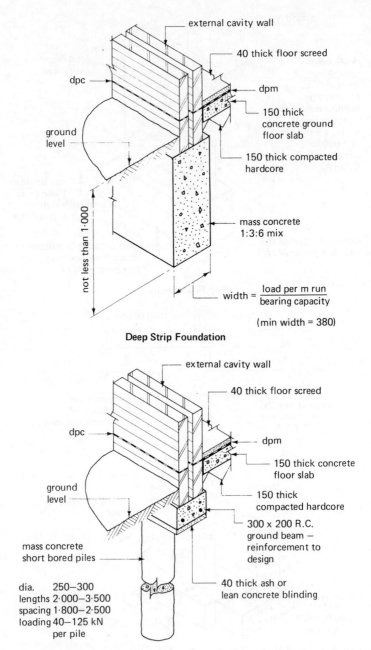

external cavity wall

40 thick floor screed

dpc

dpm

150 thick concrete ground floor slab

ground level

150 thick compacted hardcore

not less than 1·000

mass concrete 1:3:6 mix

width = $\dfrac{\text{load per m run}}{\text{bearing capacity}}$

(min width = 380)

Deep Strip Foundation

external cavity wall

40 thick floor screed

dpc

dpm

150 thick concrete floor slab

ground level

150 thick compacted hardcore

300 x 200 R.C. ground beam — reinforcement to design

mass concrete short bored piles

40 thick ash or lean concrete blinding

dia.   250–300
lengths 2·000–3·500
spacing 1·800–2·500
loading 40–125 kN
          per pile

Fig. I.11 Alternative foundations for clay soils

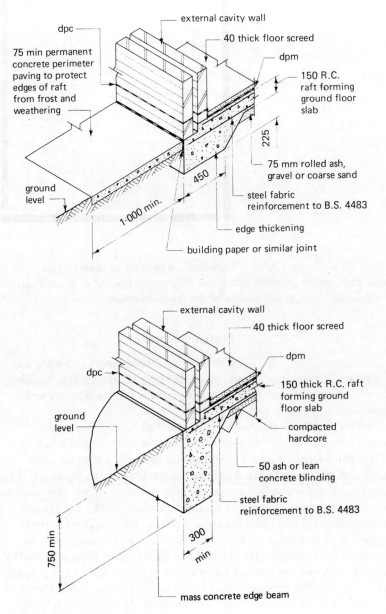

external cavity wall

dpc

75 min permanent concrete perimeter paving to protect edges of raft from frost and weathering

40 thick floor screed

dpm

150 R.C. raft forming ground floor slab

225

75 mm rolled ash, gravel or coarse sand

steel fabric reinforcement to B.S. 4483

ground level

450

1·000 min.

edge thickening

building paper or similar joint

external cavity wall

40 thick floor screed

dpc

dpm

150 thick R.C. raft forming ground floor slab

compacted hardcore

ground level

50 ash or lean concrete blinding

steel fabric reinforcement to B.S. 4483

750 min

300 min

mass concrete edge beam

Fig. I.12 Typical raft foundations

4
Concrete

Concrete is a mixture of cement, fine aggregate, coarse aggregate and water. The proportions of each material control the strength and quality of the resultant concrete.

Portland cement

Cement is the setting agent of concrete and the bulk of cement used in this country is Portland cement. This is made from chalk or limestone and clay and is generally produced by the wet process.

In this process the two raw materials are washed, broken up and mixed with water to form a slurry. This slurry is then pumped into a steel rotary kiln which is from 3-4 m in diameter and up to 150 m long and lined with refractory bricks. While the slurry is fed into the top end of the kiln a pulverised coal is blown in at the bottom end and fired. This raises the temperature at the lower end of the kiln to about $1\,400°C$. The slurry passing down the kiln first gives up its moisture, then the chalk or limestone is broken down into carbon dioxide and lime, and finally forms a white hot clinker which is transferred to a cooler before being ground. The grinding is carried out in a ball mill which is a cylinder some 15 m long and up to 4·5 m in diameter containing a large number of steel balls of various sizes, which grind the clinker into a fine powder. As the clinker is being fed into the ball mill gypsum (about 5%) is added to prevent a flash setting off the cement.

The alternative method for the preparation of Portland cement is the dry process. The main difference between this and the wet process is the reduction in the amount of water that has to be driven off in the kiln. A mixture of limestone and shale is used which is proportioned, ground and blended to form a raw meal of low moisture content. The meal is granulated in rotating pans with a small amount of water before being passed to a grate for preheating prior to entering the kiln. The kiln is smaller than that used in the wet process but its function is the same, that is, to form a clinker, which is then cooled, ground and mixed with a little gypsum as described for the previous process.

Rapid-hardening Portland cement is more finely ground than ordinary Portland cement. Its main advantage is that it gains its working strength earlier than ordinary cement. The requirements for both ordinary Portland and rapid-hardening Portland cement are given in BS 12.

High alumina cement

This is made by firing limestone and bauxite (aluminium ore) to a molten state, casting it into pigs and finally grinding it into a fine powder. Its rate of hardening is very rapid and produces a concrete which is resistant to the natural sulphates found in some subsoils. It can, however, cost up to two and a half times as much as ordinary Portland cement. The requirements of this type of cement are covered in BS 915.

Other forms of cement available are as follows:

Portland blast-furnace—BS 146.
Sulphate-resisting—BS 4027.
Low heat blast-furnace—BS 4246.
Supersulphated—BS 4248.

Cement should be stored on a damp-proof floor in the dry and kept for short periods only because eventually cement will harden as a result of the action of moisture in the air. This is known as air hardening and any hardened cement should be discarded.

Aggregates

These are the materials which are mixed with the cement to form concrete and are classed as a fine or coarse aggregate. Fine aggregates are those which will pass a standard 5 mm sieve and coarse aggregates are those which are retained on a standard 5 mm sieve. All-in aggregate is a material composed of both fine and coarse aggregates.

A wide variety of materials (for example, gravel, crushed stone, brick,

21

furnace slag and lightweight substances, such as foamed slag, expanded clay and vermiculite) are available as aggregates for the making of concrete.

In making concrete aggregates must be graded so that the smaller particles of the fine aggregate fill the voids created by the coarse aggregate. The cement paste fills the voids in the fine aggregate thus forming a dense mix.

Aggregates from natural sources are covered in BS 882.

Water

The water used in the making of concrete must be clean and free from impurities which could affect the concrete. It is usually specified as being of a quality fit for drinking. A proportion of the water will set up a chemical reaction which will harden the cement. The remainder is required to give the mix workability and will evaporate from the mix while it is curing, leaving minute voids. An excess of water will give a porous concrete of reduced durability and strength.

The quantity of water to be used in the mix is usually expressed in terms of the water/cement ratio which is:

$$\frac{\text{the total weight of water in the concrete}}{\text{weight of cement}}$$

for most mixes the ratio is between 0·4 and 0·7.

Concrete mixes can be expressed as volume ratios thus:

1 : 2 : 4 = 1 part cement, 2 parts fine aggregate and 4 parts coarse aggregate.

1 : 5 = 1 part cement and 5 parts all-in aggregate.

SOME COMMON MIXES

1 : 10—not a strong mix but it is suitable for filling weak pockets in excavations and for blinding layers.

1 : 8—slightly better than the last, suitable for paths and pavings.

1 : 6—a strong mix suitable for mass concrete foundations, paths and pavings.

1 : 3 : 6—the weakest mix equivalent to that quoted in Approved Document A as deemed to satisfy the requirements of Building Regulation A1.

1 : 2 : 4—strong mix which is practically impervious to water, in common use especially for reinforced concrete.

MIXING CONCRETE

Concrete can be mixed or batched by two methods:

1. By volume.
2. By mass.

One bag of cement has a volume of approximately 0.04 m^3 and a mass of 50 kg.

Batching by volume

This method is usually carried out using an open bottom box (of such dimensions as to make manual handling possible) called a 'gauge box'. For a 1 : 2 : 4 mix a gauge box is filled once with cement, twice with fine aggregate and four times with coarse aggregate, the top of the gauge box being struck off level each time.

If the fine aggregate is damp or wet its volume will increase by up to 25% and therefore the amount of fine aggregate should be increased by this amount. This increase in volume is called bulking.

Batching by mass

This method involves the use of a balance which is linked to a dial giving the exact mass of the materials as they are placed in the scales. This is the best method since it has a greater accuracy and the balance can be attached to the mixing machine.

Hand mixing

This should be carried out on a clean hard surface. The materials should be thoroughly mixed in the dry state before the water is added. The water should be added slowly, preferably using a rose head until a uniform colour is obtained.

Machine mixing

The mix should be turned over in the mixer for at least two minutes after adding the water. The first batch from the mixer tends to be harsh since some of the mix will adhere to the sides of the drum. This batch should be used for some less important work such as filling in weak pockets in the bottom of the excavation.

Handling

If concrete is to be transported for some distance over rough ground the runs should be kept as short as possible since vibrations of this nature can cause segregation of the materials in the mix. For the same reason concrete should not be dropped from a height of more than 1 m. If this is unavoidable a chute should be used.

Placing

If the concrete is to be placed in a foundation trench it will be levelled from peg to peg (see Fig. I.4) or if it is to be used as an oversite bed the external walls could act as a levelling guide. The levelling is carried out by tamping with a straight edge board; this tamping serves the dual purpose of compacting and bringing the excess water to the surface so that it can evaporate. Concrete must not be over tamped as this will not only bring the water to the surface but also the cement paste which is required to act as the matrix. Concrete should be placed as soon as possible after mixing to ensure that the setting action has not commenced. Concrete which dries out too quickly will not develop its full strength, therefore new concrete should be protected from the drying winds and sun by being covered with canvas, straw, polythene sheeting or damp sawdust. This protection should be continued for at least three days since concrete takes about twenty-eight days to obtain its working strength.

5

Subsoil drainage

Building Regulation C3

Wherever the dampness or position of the site of a building renders it necessary, the subsoil of the site shall be effectively drained or such other steps shall be taken as will effectively protect the building against damage from moisture.

The ideal site (see Fig. I.13) will not require any treatment but sites with a high water table will require some form of subsoil drainage. The water table is the level at which water occurs naturally below the ground and this level will vary with the seasonal changes.

The object of subsoil drainage is to lower the water table to a level such that it will comply with the above building regulation. It also has the advantage of improving the stability of the ground, lowering the humidity of the site and improving its horticultural properties.

MATERIALS

The pipes used in subsoil drainage are usually dry jointed and are either porous or perforated pipes. The porous pipes absorb the water through their walls and thus keep out the fine particles of soil or silt, whereas perforated pipes, which are laid with the perforations at the base, allow the water to rise into the pipe leaving any silt behind.

Suitable pipes
Perforated glazed-ware pipes—BS 65.

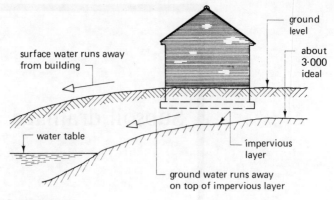

surface water runs away
from building

ground level

about 3·000 ideal

water table

impervious layer

ground water runs away
on top of impervious layer

Fig. I.13 The ideal site

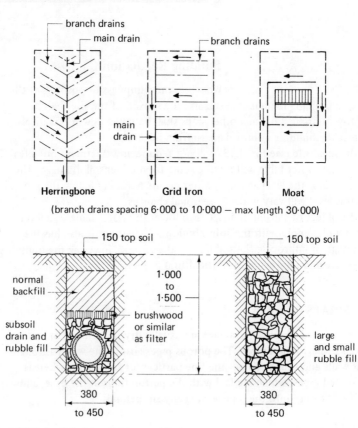

branch drains

main drain

branch drains

main drain

Herringbone   **Grid Iron**   **Moat**

(branch drains spacing 6·000 to 10·000 — max length 30·000)

150 top soil

150 top soil

normal backfill

1·000 to 1·500

subsoil drain and rubble fill

brushwood or similar as filter

large and small rubble fill

380 to 450

380 to 450

Fig. I.14 Subsoil drainage systems and drains

26

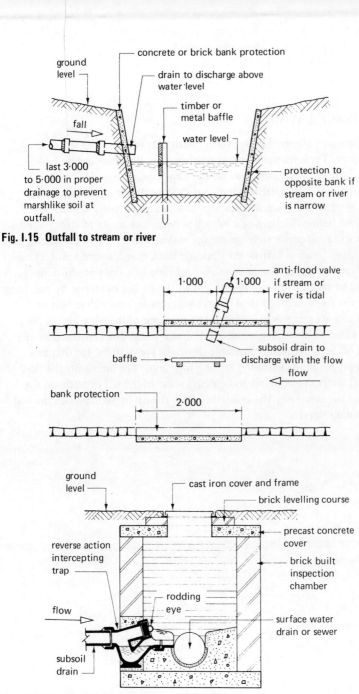

Fig. I.15 Outfall to stream or river

Fig. I.16 Outfall to surface water sewer or drain

Concrete porous pipes—BS 1194.
Clayware field pipes—BS 1196.
Perforated pitchfibre pipes—BS 2760.

DRAINAGE LAYOUTS

The pipes are arranged in a pattern to cover as much of the site as is necessary. Typical arrangements are shown on the plans in Fig. I.14. Water will naturally flow towards the easy passage provided by the drainage runs. The system is terminated at a suitable outfall such as a river, stream or surface water sewer. In all cases permission must be obtained before discharging a subsoil system. The banks of streams and rivers will need protection against the turbulence set up by the discharge and if the stream is narrow the opposite bank may also need protection (see Fig. I.15). If discharge is into a tidal river or stream precautions should be taken to ensure the system will not work in reverse by providing an outlet for the rising tide. On large schemes sediment chambers or catch pits are sometimes included to trap some of the silt which is the chief cause of blockages in subsoil drainage work. The construction of a catch pit is similar to the manhole shown in Fig. I.16 except that in a catch pit the inlet and outlet are at a high level, this interrupts the flow of subsoil water in the drains and enables some of the silt to settle on the base of the catch pit. The collected silt in the catch pit must be removed at regular intervals.

Part II
Superstructure

Part II

Superstructure

6

Stonework, brickwork and blockwork

STONEWORK

The natural formation of stones or rocks is a very long process which commenced when the earth was composed only of gases. These gases eventually began to liquefy forming the land and the sea, the land being only a thin crust or mantle to the still molten core.

Igneous rocks originated from the molten state; plutonic rocks are those in the lower part of the earth's mantle; whereas hypabyssal rocks solidified rapidly near the upper surface of the crust.

Building stones

Stones used in building can be divided into three classes as follows:

1. Igneous.
2. Sedimentary.
3. Metamorphic.

IGNEOUS STONES

These stones originate from volcanic action being formed by the crystallisation of molten rock matter derived from deep in the earth's crust. It is the proportions of these crystals which give the stones their colour and characteristics. Granites are typical of this class of stone being hard, durable and capable of a fine polished finish. Granites are mainly composed of quartz, felspar and mica.

SEDIMENTARY STONES

These stones are largely composed of material derived from the breakdown and erosion of existing rocks deposited in layers under the waters, which at that time covered much of the earth's surface. Being deposited in this manner their section is stratified and shows to a lesser or greater degree the layers as deposited. Some of these layers are only visible if viewed under a microscope. Under the microscope it will be seen that all the particles lie in one direction indicating the flow of the current in the water. Sandstones and limestones are typical examples of sedimentary stones.

Sandstones are stratified sedimentary rocks produced by the eroded and disintegrated rocks, like granite, being carried away and deposited by water in layers. The brown and yellow tints in sandstones are due to the presence of oxides of iron.

Limestones may be organically formed by the deposit of tiny shells and calcareous (containing limestone) skeletons in the seas and rivers or it may be formed chemically by deposits of lime in ringed layers. Limestones vary considerably from a heavy crystalline form to a friable material such as chalk.

METAMORPHIC STONES

These are stones which have altered and may have been originally igneous or sedimentary rocks, but have since been changed by geological processes such as pressure, movement, heat and chemical reaction due to the infiltration of fluids. Typical examples of this type of stone are marbles and slates.

Marbles are metamorphic limestones, their original structure having been changed by pressure. Marbles being capable of taking a high polish are used mainly for decorative work.

Slate is a metamorphic clay having been subjected to great pressure and heat; being derived from a sedimentary layer it can be easily split into thin members.

Stones are obtained from quarries by blasting and wedging the blocks away from the solid mass. They are partly worked in the quarry and then sent to store yards where they can be sawn, cut, moulded, dressed and polished to the customer's requirements.

Today, natural stones are sometimes used for facing prestige buildings, constructing boundary or similar walls and in those areas where natural

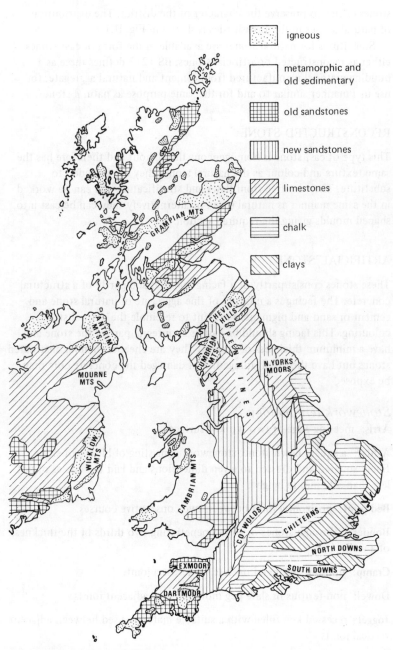

Legend:

- igneous
- metamorphic and old sedimentary
- old sandstones
- new sandstones
- limestones
- chalk
- clays

Fig. II.1 Distribution of natural stones

Map labels:
GRAMPIAN MTS, ANTRIM, MOURNE MTS, CHEVIOT HILLS, CUMBRIAN MTS, PENNINES, N. YORKS MOORS, WICKLOW MTS, CAMBRIAN MTS, COTSWOLDS, CHILTERNS, NORTH DOWNS, SOUTH DOWNS, EXMOOR, DARTMOOR

stones occur, to preserve the character of the district. The distribution of natural stones in the British Isles is shown in Fig. II.1.

Substitutes for natural stones are available in the form of cast stones either as reconstructed or artificial stones. BS 1217 defines these as a building material manufactured from cement and natural aggregate, for use in a manner similar to and for the same purpose as natural stone.

RECONSTRUCTED STONE

This type of cast stone is homogeneous throughout and therefore has the same texture and colour as the natural stones they are intended to substitute. They are free from flaws and stratification and can be worked in the same manner as natural stone or alternatively they can be cast into shaped moulds giving the required section.

ARTIFICIAL STONE

These stones consist partly of a facing material and partly of a structural concrete. The facing is a mixture of fine aggregate of natural stone and cement or sand and pigmented cement to resemble the natural stone colouring. This facing should be cast as an integral part of the stone and have a minimum thickness of 20 mm. They are cheaper than reconstructed stones but have the disadvantage that if damaged the concrete core may be exposed.

Stonework terminology

Arris: meeting edge of two worked surfaces.

Ashlar: a square hewn stone; stonework consisting of blocks of stone finely squared and dressed to given dimensions and laid to courses of not less than 300 mm in height.

Bed joint: horizontal joint between two consecutive courses.

Bonders: through stones or stones penetrating two thirds of the thickness of a wall.

Cramp: non-ferrous metal or slate tie across a joint.

Dowel: non-ferrous or slate peg morticed into adjacent joints.

Joggel: recessed key filled with a suitable material, used between adjacent vertical joints.

Lacing: course of different material to add strength.

Natural bed: plane of stratification in sedimentary stones.

34

Quarry sap: moisture contained in newly quarried stones.

Quoin: corner stone.

Stool: flat seating on a weathered sill for jamb or mullion.

String course: distinctive course or band used mainly for decoration.

Weathering: sloping surface to part of the structure to help shed the rain.

ASHLAR WALLING

This form of stone walling is composed of carefully worked stones, regularly coursed, bonded and set with thin or rusticated joints and is used for the majority of high-class facing work in stone. The quions are sometimes given a surface treatment to emphasise the opening or corner of the building. The majority of ashlar work is carried out in limestone varying in thickness from 100-300 mm and set in mason's putty which is a mixture of stone dust, lime putty and portland cement a typical mix ratio being 7 : 5 : 2.

Rules for ashlar work
1. Back faces of ashlar stones should be painted with a bituminous or similar waterproofing paint.
2. External stonework must not be taken through the thickness of the wall since this could create a passage for moisture.
3. Ledges of cornices and external projections should be covered with lead, copper or asphalt to prevent damage by rain or birds.
4. Moulded cornices should be raked back at 45° to counteract the cantilever action.
5. Face of stones should be given a protective coat of slurry during construction, the slurry being washed off immediately prior to completion.

Typical details of ashlar work are shown in Figs. II.2-II.5.

RUBBLE WALLING

These are walls consisting of stones which are left in a rough or uneven state thus presenting a natural appearance to the face of the wall. These stones are usually laid with a wide joint and are frequently used in various forms in many rural areas. They can be laid dry or bedded in earth in boundary walls or bedded in lime mortar when used for the walls of farm outbuildings, if used in conjunction with ashlar stonework a cement or gauged mortar is used. It is usual to build the quoins to corners, window

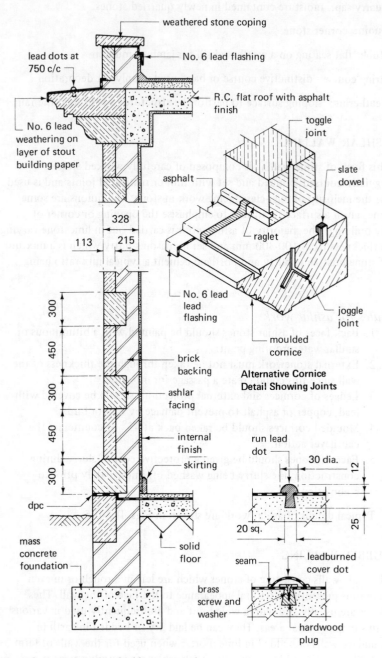

Fig. II.2 Typical details of ashlar stonework

weathered stone coping

lead dots at 750 c/c

No. 6 lead flashing

No. 6 lead weathering on layer of stout building paper

R.C. flat roof with asphalt finish

asphalt

toggle joint

slate dowel

raglet

joggle joint

No. 6 lead lead flashing

moulded cornice

Alternative Cornice Detail Showing Joints

328

113   215

300

450

300

450

300

brick backing

ashlar facing

internal finish

skirting

run lead dot

30 dia.

12

20 sq.   25

dpc

mass concrete foundation

solid floor

seam

leadburned cover dot

brass screw and washer

hardwood plug

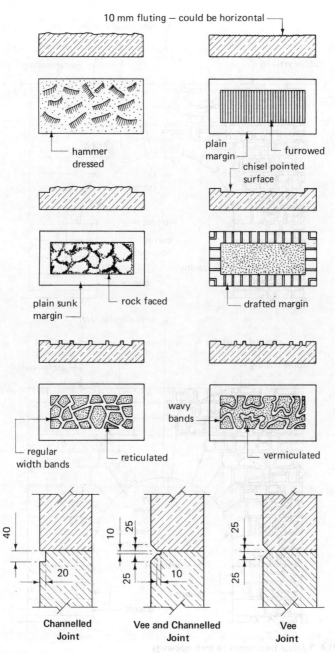

Fig. II.3 Typical surface and joint treatments

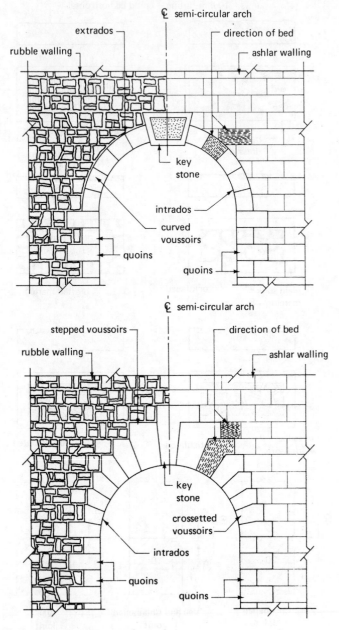

Fig. II.4 Typical treatments to arch openings

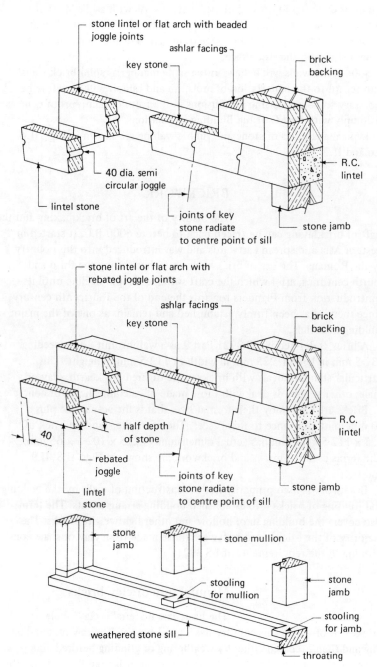

Fig. II.5 Typical treatments to square openings

Labels in figure:
- stone lintel or flat arch with beaded joggle joints
- ashlar facings
- key stone
- brick backing
- 40 dia. semi circular joggle
- R.C. lintel
- lintel stone
- joints of key stone radiate to centre point of sill
- stone jamb
- stone lintel or flat arch with rebated joggle joints
- ashlar facings
- key stone
- brick backing
- R.C. lintel
- half depth of stone
- 40
- rebated joggle
- lintel stone
- joints of key stone radiate to centre point of sill
- stone jamb
- stone jamb
- stone mullion
- stooling for mullion
- stone jamb
- stooling for jamb
- weathered stone sill
- throating

and door openings in dressed or ashlar stones. As with ashlar work it is advisable to treat the face of any backing material with a suitable water-proofing coat to prevent the passage of moisture or the appearance of cement stains on the stone face.

Solid stone walls will behave in the same manner as solid brick walls with regard to the penetration of moisture and rain, it will therefore be necessary to take the same precautions in the form of damp-proof courses to comply with Part C of the Building Regulations.

Typical examples of stone and rubble walling are shown in Figs. II.4, II.6 and II.7.

BRICKWORK

The history of the art of brickmaking and the craft of bricklaying can be traced back to before 6000 B.C. It started in western Asia and spread eastwards and was introduced into this country by the Romans. The use of brickwork flourished during the third and fourth centuries, after which the craft suffered a rapid decline until its reintroduction from Flanders towards the end of the fourteenth century. Since then it has been firmly established and remains as one of the major building materials.

A brick is defined in BS 3921 Part 2 as a walling unit not exceeding 337·5 mm in length, 225 mm in width or 112·5 mm in height. This particular standard deals with bricks made of fired brickearth, clay or shale; other standards deal with those made of calcium silicate or concrete.

Bricks are known by their format size, that is the actual size plus a 10 mm joint allowance to three faces. Therefore the standard brick of 225 x 112·5 x 75 mm has actual dimensions of 215 x 102·5 x 65 mm. The terms used for bricks and brickwork are shown in Figs. II.8, II.9 and II.10.

Brickwork is used primarily in the construction of walls by the bedding and jointing of bricks into established bonding arrangements. The term also covers the building in of hollow and other lightweight blocks. The majority of the bricks used today are those made from clay or shale conforming to the requirements of BS 3921.

Manufacture of clay bricks

The basic raw material is clay, shale or brickearth of which this country has a good supply. The raw material is dug and then prepared either by weathering or grinding before being mixed with water to the right plastic condition. It is then formed into the required brick shape before being dried and fired in a kiln.

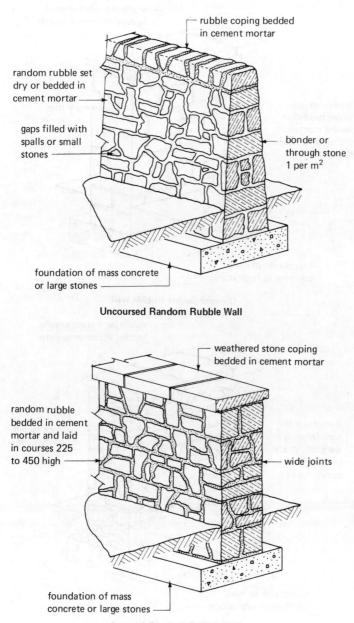

rubble coping bedded in cement mortar

random rubble set dry or bedded in cement mortar

gaps filled with spalls or small stones

bonder or through stone 1 per m²

foundation of mass concrete or large stones

Uncoursed Random Rubble Wall

weathered stone coping bedded in cement mortar

random rubble bedded in cement mortar and laid in courses 225 to 450 high

wide joints

foundation of mass concrete or large stones

Coursed Random Rubble Wall

Fig. II.6 Typical rubble walls

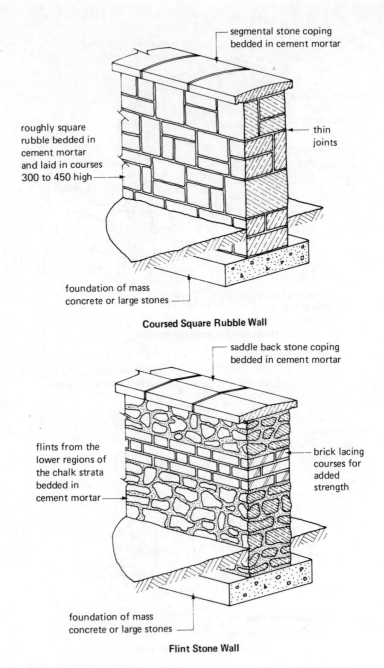

segmental stone coping bedded in cement mortar

thin joints

roughly square rubble bedded in cement mortar and laid in courses 300 to 450 high

foundation of mass concrete or large stones

Coursed Square Rubble Wall

saddle back stone coping bedded in cement mortar

brick lacing courses for added strength

flints from the lower regions of the chalk strata bedded in cement mortar

foundation of mass concrete or large stones

Flint Stone Wall

Fig. II.7 Typical stone walls

42

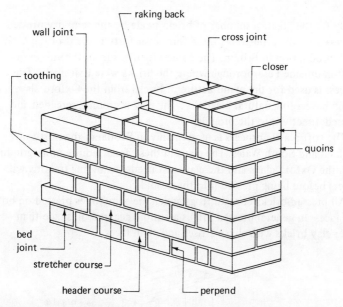

Fig. II.8 Brickwork terminology

Different clays have different characteristics, such as moisture content and chemical composition, therefore distinct variations of the broad manufacturing processes have been developed and these are easily recognised by the finished product.

PRESSED BRICKS

This type of brick is the most common used accounting for nearly two-thirds of the 8 400 million produced in this country each year. There are two processes of pressed brick manufacture they are the semi-dry and stiff plastic methods.

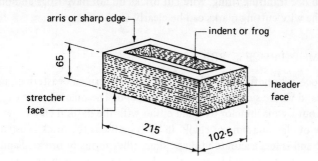

Fig. II.9 Standard brick

By far the greatest number of bricks made by the semi-dry pressed process are called 'flettons', these form over forty percent of the total brick production in Britain. The name originates from the village of Fletton outside Peterborough where the bricks were first made. This process is used for the manufacture of bricks from the Oxford clays which have a low natural plasticity. The clay is ground, screened and pressed directly into the moulds.

The stiff plastic process is used mainly in Scotland, the north of England and South Wales. The clays in these areas require more grinding than the Oxford clays and the clay dust needs tempering (mixing with water) before being pressed into the mould.

All pressed bricks contain frogs which are sometimes pressed on both bed faces. In general pressed bricks are more accurate in shape than other clay bricks with sharp arrises and plain faces.

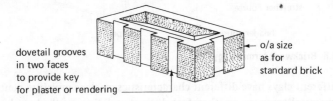

dovetail grooves
in two faces
to provide key
for plaster or rendering

o/a size
as for
standard brick

Fig. II.10 Keyed brick

WIRE CUT BRICKS

Approximately 28% of bricks produced in Britain are made by this process. The clay which is usually fairly soft and of a fine texture is extruded as a continuous ribbon and is cut into brick units by tightly stretched wires spaced at the height or depth for the required brick. Allowance is made during the extrusion and cutting for the shrinkage that will occur during firing. Wire cut bricks do not have frogs and on many the wire cutting marks can be clearly seen.

SOFT MUD PROCESS BRICKS

This process is confined mainly to the south-eastern and eastern counties of England where suitable soft clays are found. The manufacture can be carried out by machine or by hand either with the natural clay or with a mixture of clay and lime or chalk. In both methods the brick is usually frogged and is less accurate in shape than other forms of bricks. Sand is usually used in the moulds, to enable the bricks to be easily removed, and this causes an uneven patterning or creasing on the face.

Brick classification

No standard system for the classification of bricks has yet been devised, bricks are generally known by the terms given in BS 3921 or by the description given by the brick manufacturer or a combination of the two.

BS 3921, PART 2

This standard gives three headings:

1. Varieties

Common: suitable for general building work but having no special claim to give an attractive appearance.

Facing: specially made or selected to have an attractive appearance when used without rendering or plaster.

Engineering: having a dense and strong semi-vitreous body conforming to defined limits for absorption and strength.

2. Qualities

Internal: suitable for internal use only, may need protection on site during bad weather or during the winter.

Ordinary: less durable than special quality but normally durable in the external face of a building. Some types are unsuitable for exposed situations.

Special: for use in conditions of extreme exposure where the structure may become saturated and frozen such as retaining walls and pavings.

3. Types

Solid: those in which small holes passing through or nearly through the brick do not exceed 25% of its volume or in which frogs do not exceed 20% of its volume. A small hole is defined as a hole less than 20 mm wide or less than 500 mm^2 in area.

Perforated: those in which holes passing through the brick exceed 25% of its volume and the holes are small as defined above.

Hollow: those in which the holes passing through the brick exceed 25% of its volume and the holes are larger than those defined as small holes.

Cellular: those in which the holes are closed at one end and exceed 20% of the volume of the brick.

45

Bricks may also be classified by one or more of the following:

Place of origin, for example, London.

Raw material, for example, clay.

Manufacture, for example, wire cut.

Use, for example, foundation.

Colour, for example, blue.

Surface texture, for example, sand-faced.

Calcium silicate bricks

These bricks are also called sandlime and sometimes flintlime bricks and are covered by BS 187, Part 2, which gives eight classes of brick, the higher the numbered class the stronger is the brick. The format size of a calcium silicate brick is the same as that given for a standard clay brick.

These bricks are made from carefully selected clean sand and/or crushed flint mixed with controlled quantities of lime and water. At this stage colouring pigments can be added if required, the relatively dry mix is then fed into presses to be formed into the required shape. The moulded bricks are then hardened in sealed and steam pressurised autoclaves. This process, which takes from seven to ten hours, causes a reaction between the sand and the lime resulting in a strong homogenous brick which is ready for immediate delivery and laying. The bricks are very accurate in size and shape but do not have the individual character of clay bricks.

Concrete bricks

These are made from a mixture of inert aggregate and cement in a similar fashion to calcium silicate bricks and are cured either by natural weathering or in an autoclave. Details of the types and properties available as standard concrete bricks are given in BS 6073.

Mortars for brickwork

The mortar used in brickwork transfers the stresses, tensile, compressive and shear uniformly between adjacent bricks. To do this it must satisfy certain requirements:

1. Have adequate strength, but not greater than that required for the design strength.
2. Have good workability.
3. Needs to retain plasticity long enough for the bricks to be laid.

4. Must be durable over a long period.
5. Bond well to the bricks.
6. Should be able to be produced at an economic cost.

If the mortar is weaker than the bricks shrinkage cracks will tend to follow the joints of the brickwork and these are reasonably easy to make good. If the mortar is stronger than the bricks shrinkage cracks will tend to be vertical through the joints and the bricks thus weakening the fabric of the structure.

MORTAR MIXES

Mortar is a mixture of sand and lime or a mixture of sand and cement with or without lime. Proportioning of the materials can be carried by volume but this method is inaccurate and it is much better to proportion by weight. The effect of the lime is to make the mix more workable, but as the lime content increases the mortar's resistance to damage by frost action decreases.

Plasticisers by having the effect of entraining small bubbles of air in the mix and breaking down surface tension will also increase the workability of a mortar.

Mortars should never be re-tempered and should be used within two hours of mixing or be discarded.

Typical mixes (by volume)

Cement mortar 1 : 3 (cement : sand) suitable for brickwork in exposed conditions such as parapets and for brickwork in foundations.

Lime mortar 1 : 3 (lime : sand) for internal use only.

Gauged mortars (cement : lime : sand):
1 : 1 : 6 suitable for most conditions of severe exposure.
1 : 2 : 9 suitable for most conditions except those of severe exposure.
1 : 3 : 12 internal use only.

Dampness penetration

It is possible for dampness to penetrate into a building through the brick walls by one or more of three ways:

1. By the rain penetrating the head of the wall and soaking down into the building below the roof level.
2. By the rain beating against the external wall and soaking through the fabric into the building.
3. By ground moisture entering the wall at or near to the base and

47

creeping up the wall by capillary action and entering the building above the ground floor level.

Nos. 1 and 3 can be overcome by the insertion of a suitable damp-proof course in the thickness of the wall. 2 can be overcome by one of two methods:

(a) Applying to the exposed face of the wall a barrier such as cement rendering or some suitable cladding like vertical tile hanging.

(b) by constructing a cavity wall, whereby only the external skin becomes damp, the cavity providing a suitable barrier to the passage of moisture through the wall.

DAMP-PROOF COURSES

The purpose of a damp-proof course in a building is to provide a barrier to the passage of moisture from an external source into the fabric of the building or from one part of the structure to another. Damp-proof courses may be either horizontal or vertical and can generally be divided into three types:

1. Those below ground level to prevent the entry of moisture from the soil.
2. Those placed just above ground level to prevent moisture creeping up the wall by capillary action, this is sometimes called rising damp.
3. Those placed at openings, parapets and similar locations to exclude the entry of the rainwater which falls directly onto fabric of the structure.

Materials for damp-proof courses

BS 743 gives seven suitable materials for the construction of damp-proof courses all of which should have the following properties:

(a) Completely impervious.

(b) Durable, having a longer life than the other components in the building and therefore should not need replacing during its life-time.

(c) Be in comparatively thin sheets so as to prevent disfigurement of the building.

(d) Be strong enough to support the loads placed upon it without exuding from the wall.

(e) Be flexible enough to give with any settlement of the building without fracturing.

Lead: this should be at least code No. 4—BS 1178 lead, it is a flexible

48

material supplied in thin sheets and therefore large irregular shapes with few joints can be formed but it has the disadvantages of being expensive and liable to exude under heavy loadings.

Copper: should have a minimum thickness of 0·25 mm, like lead it is supplied in thin sheets and is expensive.

Bitumen: this is supplied in the form of a felt usually to brick widths and is therefore laid quickly with the minimum number of joints. Various bases are available, such as hessian, fibre, asbestos and lead, all of which are inexpensive but have the main disadvantage of being easily torn.

Mastic asphalt: applied in two layers giving a total thickness of 25 mm, it is applied *in situ* and is therefore jointless but is expensive in small quantities.

Polythene: black low density polythene sheet of single thickness not less than 0·5 mm thick should be used, it is easily laid but can be torn and punctured easily.

Slates: these should not be less than 230 mm long nor less than 4 mm thick and laid in two courses set breaking the joint in cement mortar 1 : 3. Slates have limited flexibility but are impervious and very durable, cost depends upon the area in which the building is being erected.

Bricks: should comply with the requirements of BS 3921 and be laid in two courses in cement mortar 1 : 3, can be out of context with the general façade of the building.

Brickwork bonding

When building with bricks it is necessary to lay the bricks to some recognised pattern or bond in order to ensure stability of the structure and to produce a pleasing appearance. All the various bonds are designed so that no vertical joint in any one course is directly above or below a vertical joint in the adjoining course. To simplify this requirement special bricks are produced or cut from whole bricks on site, a selection of these special bricks is shown in Fig. II.11. The various bonds are also planned to give the greatest practical amount of lap to all the bricks and this should not be less than a quarter of a brick length. Properly bonded brickwork distributes the load over as large an area of brickwork as possible, then angle of spread of the load through the bonded brickwork is 60°.

Common bonds
Stretcher bond: consists of all stretchers in every course and is used for

half brick walls and the half brick skins of hollow or cavity walls (see Fig. II.12).

English bond: a very strong bond consisting of alternate courses of headers and stretchers (see Fig. II.13).

Flemish bond: each course consists of alternate headers and stretchers, its appearance is considered to be better than English bond but it is not quite so strong. This bond requires fewer facing bricks than English bond needing only 79 bricks per square metre as opposed to 89 facing bricks per square metre for English bond. This bond is sometimes referred to as double Flemish bond (see Fig. II.14).

Single Flemish bond: a combination of English and Flemish bonds having Flemish bond on the front face with a backing of English bond. It is considered to be slightly stronger than Flemish bond. The thinnest wall that can be built using this bond is a one and half brick wall.

English garden wall bond: consists of three courses of stretchers to one course of headers.

Flemish garden wall bond: consists of one header to every three stretchers in every course, this bond is fairly economical in facing bricks and has a pleasing appearance.

Special bonds

Rat-trap bond: this is a brick on edge bond and gives a saving on materials and loadings, suitable as a backing wall to a cladding such as tile hanging (see Fig. II.15).

Quetta bond: used on one and half brick walls for added strength, suitable for retaining walls (see Fig. II.15).

METRIC MODULAR BRICKWORK

The standard format brick does not fit reasonably well into the system of dimensional coordination with its preferred dimension of 300 mm, therefore metric modular bricks have been designed and produced with four different formats (see Fig. II.16).

The bond arrangements are similar to the well-known bonds but are based on third bonding, that is the overlap is one-third of a brick and not one-quarter as with the standard format brick. Examples of metric modular brick bonding are shown in Fig. II.16.

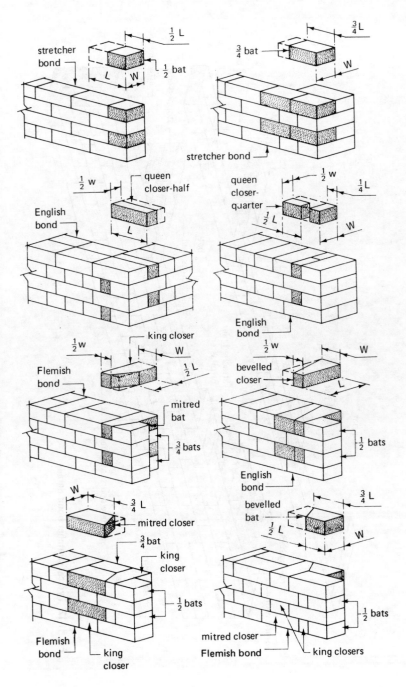

Fig. II.11 Special bricks

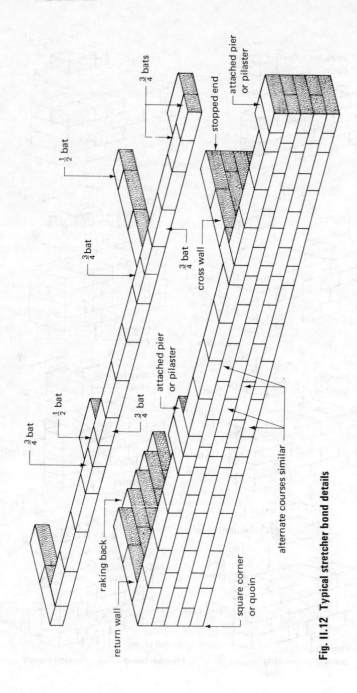

Fig. II.12 Typical stretcher bond details

½ bat

¾ bats

¾ bat

attached pier or pilaster

stopped end

cross wall

¾ bat

½ bat

¾ bat

attached pier or pilaster

raking back

square corner or quoin

alternate courses similar

return wall

52

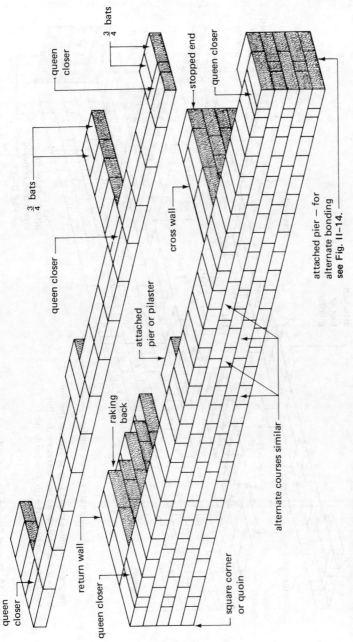

Fig. II.13 Typical English bond details

queen closer

$\frac{3}{4}$ bats

queen closer

$\frac{3}{4}$ bats

queen closer

stopped end

queen closer

cross wall

attached pier or pilaster

raking back

return wall

queen closer

attached pier — for alternate bonding see Fig. II-14.

alternate courses similar

square corner or quoin

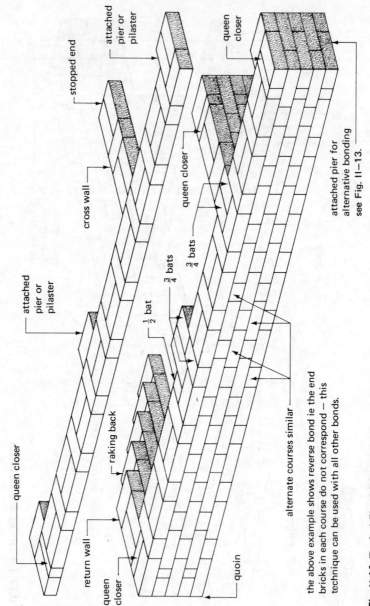

54

Fig. II.14 Typical Flemish bond details

attached pier or pilaster

queen closer

stopped end

cross wall

attached pier or pilaster

queen closer

queen closer

attached pier for alternative bonding see Fig. II—13.

raking back

½ bat

¾ bats

¾ bats

return wall

queen closer

quoin

alternate courses similar

the above example shows reverse bond ie the end bricks in each course do not correspond — this technique can be used with all other bonds.

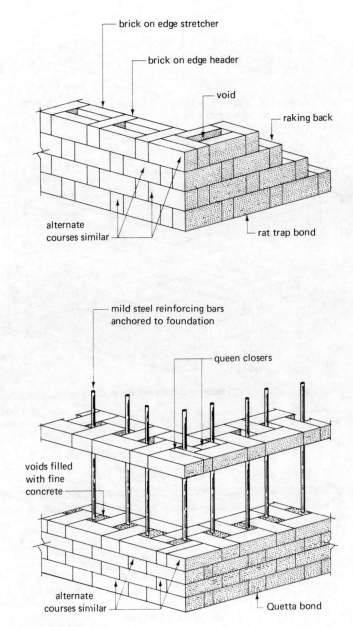

brick on edge stretcher

brick on edge header

void

raking back

alternate
courses similar

rat trap bond

mild steel reinforcing bars
anchored to foundation

queen closers

voids filled
with fine
concrete

alternate
courses similar

Quetta bond

Fig. II.15 Special bonds

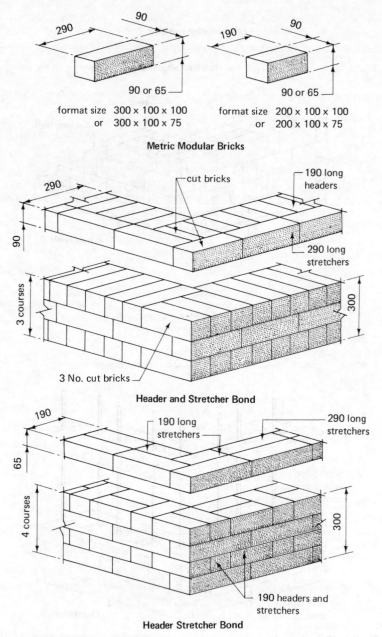

format size 300 × 100 × 100
or 300 × 100 × 75

format size 200 × 100 × 100
or 200 × 100 × 75

Metric Modular Bricks

cut bricks

190 long headers

290 long stretchers

3 No. cut bricks

Header and Stretcher Bond

190 long stretchers

290 long stretchers

190 headers and stretchers

Header Stretcher Bond

Fig. II.16 Metric modular brickwork

FOOTINGS

These are wide courses of bricks placed at the base of a wall to spread the load over a greater area of the foundations. This method is seldom used today, instead the concrete foundation would be reinforced to act as a beam or reinforced strip. The courses in footings are always laid as headers as far as possible, stretchers if needed are laid in the centre of the wall (see Fig. II.17).

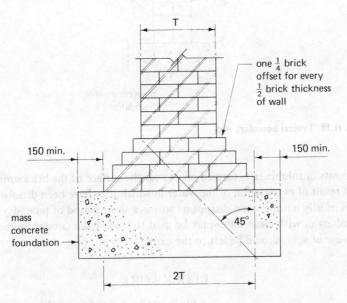

T

one $\frac{1}{4}$ brick offset for every $\frac{1}{2}$ brick thickness of wall

150 min.

150 min.

mass concrete foundation

45°

2T

Fig. II.17 Typical footings

BOUNDARY WALLS

These are subjected to severe weather conditions and therefore should be correctly designed and constructed. If these walls are also acting as a retaining wall the conditions will be even more extreme, but the main design principle of the exclusion of water remains the same. The presence of water in brickwork can lead to frost damage, mortar failure and efflorescence. The incorporation of adequate damp-proof courses and overhanging throated copings is of the utmost importance in this form of structure (see Fig. II.18).

Efflorescence

This is a white stain appearing on the face of brickwork caused by

57

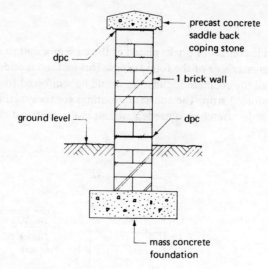

precast concrete
saddle back
coping stone

dpc

1 brick wall

ground level

dpc

mass concrete
foundation

Fig. II.18 Typical boundary wall

deposits of soluble salts formed on or near the surface of the brickwork as a result of evaporation of the water in which they have been dissolved. It is usually harmless and disappears within a short period of time; dry brushing or with clean water may be used to remove the salt deposit but the use of acids should be left to the expert.

BLOCKWORK

A block can be defined as a walling unit exceeding the dimensions specified for bricks given in BS 3921 and that its height shall not exceed either its length or six times its thickness to avoid confusion with slabs or panels. Blocks are produced from clay, precast concrete and aerated concrete.

Clay Blocks

These are covered by BS 3921 which gives a format size of 300 x 225 x 62·5; 75, 100 or 150 mm wide. These blocks, which are hollow, are made by an extrusion process and fired as for clay bricks. The standard six cavity block is used mainly for the inner skin of a cavity wall, whereas the three cavity block is primarily intended for partition work. Special corner, closer, fixing and conduit blocks are produced to give the range good flexibility in design and lay-out. Typical details are shown in Fig. II.19.

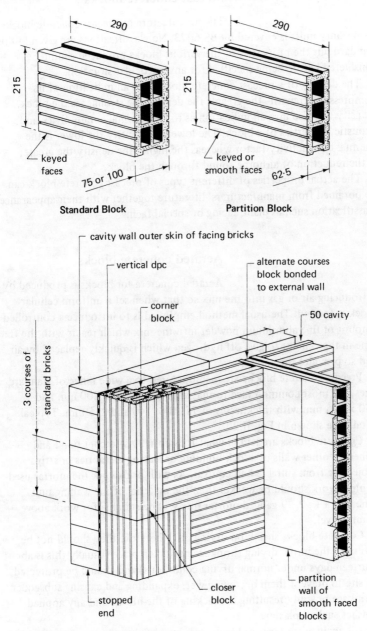

Fig. II.19 Hollow clay blocks and blockwork

The following labels appear in the figure:

Standard Block
- 290
- 215
- keyed faces
- 75 or 100

Partition Block
- 290
- 215
- keyed or smooth faces
- 62·5

Blockwork:
- cavity wall outer skin of facing bricks
- vertical dpc
- alternate courses block bonded to external wall
- corner block
- 50 cavity
- 3 courses of standard bricks
- closer block
- stopped end
- partition wall of smooth faced blocks

59

Precast concrete blocks

The manufacture of Precast concrete blocks or masonry units is covered by BS 6073. No classifications are given in this standard but the properties of the various blocks produced should be considered before being specified for any particular situation.

The density of a precast concrete block gives an indication of its compressive strength—the greater the density the stronger is the block. Density will also give an indication as to the thermal conductivity and acoustic properties of a block. The lower the density the lower is the thermal conductivity factor whereas the higher the density the greater is the reduction of airborne sound through the block.

The actual properties of different types of precast concrete block can be obtained from manufacturers' literature together with their appearance classification such as plain, facing or special facing.

Aerated concrete blocks

Aerated concrete for blocks is produced by introducing air or gas into the mix so that when set a uniform cellular block is formed. The usual method employed is to introduce a controlled amount of fine aluminium powder into the mix which reacts with the free lime in the cement to give off hydrogen which is quickly replaced by air and so provides the aeration.

Precast concrete blocks are manufactured to a wide range of standard sizes, the most common face format sizes being 400 x 200 mm and 450 x 225 mm with thicknesses of 75, 100, 140 and 215 mm. Typical details are shown in Fig. II.20.

Concrete blocks are laid in what is essentially stretcher bond and joined to other walls by block bonding or leaving metal ties or strips projecting from suitable bed courses. As with brickwork the mortar used in blockwork should be weaker than the material of the walling unit, generally a 1 : 2 : 9 gauged mortar mix will be suitable for work above ground level.

Concrete blocks shrink on drying out, therefore they should not be laid until the initial drying shrinkage has taken place (usually this is about fourteen days under normal drying conditions) and should be protected on site to prevent them becoming wet, expanding and causing subsequent shrinkage possibly resulting in cracking of the blocks and any applied finishes such as plaster.

The main advantages of blockwork over brickwork are:

1. Labour saving—easy to cut, larger units.

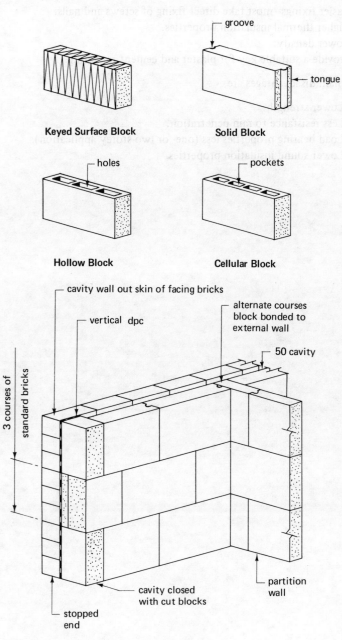

Fig. 11.20 Precast concrete blocks and blockwork

Keyed Surface Block

Solid Block

groove

tongue

Hollow Block

holes

Cellular Block

pockets

cavity wall out skin of facing bricks

vertical dpc

alternate courses block bonded to external wall

50 cavity

3 courses of standard bricks

cavity closed with cut blocks

partition wall

stopped end

2. Easier fixings—most take direct fixing of screws and nails.
3. Higher thermal insulation properties.
4. Lower density.
5. Provide a suitable key for plaster and cement rendering.

The main disadvantages are:

(*a*) Lower strength.
(*b*) Less resistance to rain penetration.
(*c*) Load bearing properties less (one- or two-storey application).
(*d*) Lower sound insulation properties.

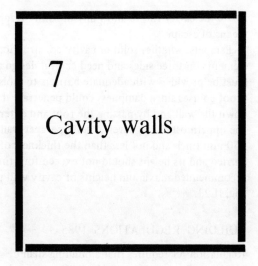

7
Cavity walls

A wall constructed in two leaves or skins with a space or cavity between them is called a cavity wall and it is the most common form of external wall used in domestic building today. The main purpose of constructing a cavity wall is to prevent the penetration of rain to the internal surface of the wall. It is essential that the cavity is not bridged in any way as this would provide a passage for the moisture.

Air bricks are sometimes used to ventilate the cavity and these would be built in at the head and base of the cavity wall in order that a flow of air may pass through the cavity thus drying out any moisture that has penetrated the outer leaf. Unless the wall is exposed to very wet conditions the practice of inserting air bricks to ventilate the cavity is not recommended since it lowers the thermal and sound insulation values of the wall.

The main consideration in the construction of a cavity wall above ground level damp-proof course is the choice of a brick or block which will give the required durability, strength and appearance and also conform to Building Regulation requirements. The main function of the wall below ground level damp-proof course is to transmit the load safely to the foundations, in this context the two half-brick leaves forming the wall act as retaining walls. There is a tendency for the two leaves to move towards each other, due to the pressure of the soil, and the space provided by the cavity. To overcome this problem it is common practice to fill the cavity below ground level with a weak mix of concrete thus creating a 'solid' wall in the ground (see Fig. II.22). It is also advisable to leave out every fourth vertical joint in the external leaf at the base of the cavity and above the cavity fill, to allow any moisture trapped in the cavity a means of escape.

Parapets, whether solid or cavity construction, are exposed to the elements on three sides and need careful design and construction. They must be provided with adequate barriers to moisture in the form of damp-proof courses since dampness could penetrate the structure by soaking down the wall and by-passing the roof and entering the building below the uppermost ceiling level. A solid parapet wall should not be less than 150 mm thick and not less than the thickness of the wall on which it is carried and its height should not exceed four times its thickness. The recommended maximum heights of cavity wall parapets is shown in Fig. II.23.

BUILDING REGULATIONS 1985

Regulation A1 requires that a building shall be so constructed that the combined dead, imposed and wind loads are sustained and transmitted to the ground safely and without causing any movement which will impair the stability of any part of another building. Guidance to meet the above requirements for cavity walls is given in Approved Document A.

Part C of this document deals with full storey height cavity walls for residential buildings of up to three storeys and recommends that:

1. The unit compressive strengths of bricks and blocks should be 5 N/mm^2 and 2.8 N/mm^2 respectively.
2. Cavities should be not less than 50 mm nor more than 100 mm in width at any level.
3. Wall ties should comply with BS 1243 or other not less suitable type and should have the maximum spacings given in Table C3. Fig. II.21 gives the maximum wall tie spacings for a cavity width of 50 to 75 mm.
4. All cavity walls should have leaves at least 90 mm thick.
5. The combined thickness of the two leaves of a cavity wall plus 10 mm should not be less than the thickness required for a solid wall of the same length and height (for a maximum wall length of 9.000 with a height between 3.500 and 9.000 the two leaves plus 10 mm is 200 mm for the whole of its height—see Table C2 in AD. A).
6. Mortar should be as given for mortar designation (iii) in BS 5628 Part 1 or a gauged mortar mix of 1 : 1 : 6 by volume.
7. Cavity walls of any length need to be provided with roof lateral support and those over 3.000 in length will also require floor lateral support at every floor forming a junction with the supported wall.

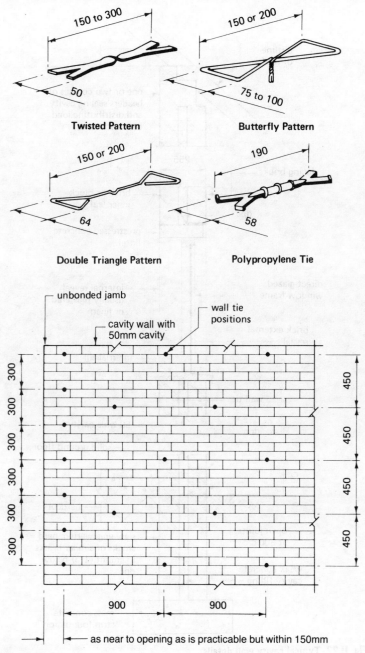

Fig. II.21 Wall ties

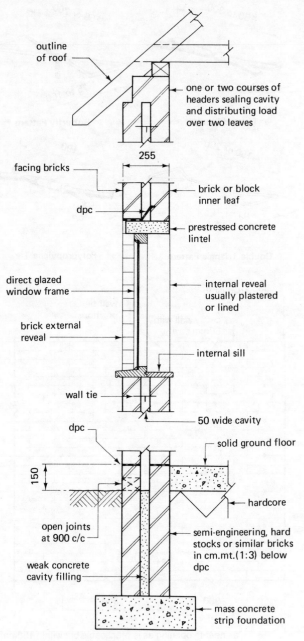

outline of roof

one or two courses of headers sealing cavity and distributing load over two leaves

255

facing bricks

brick or block inner leaf

dpc

prestressed concrete lintel

direct glazed window frame

internal reveal usually plastered or lined

brick external reveal

internal sill

wall tie

50 wide cavity

dpc

solid ground floor

150

hardcore

open joints at 900 c/c

semi-engineering, hard stocks or similar bricks in cm.mt.(1:3) below dpc

weak concrete cavity filling

mass concrete strip foundation

Fig. II.22 Typical cavity wall details

If roof lateral support is not provided by type of covering (tiles or slates), a pitch of 15° or more plus a minimum bearing of 75 mm then durable metal straps with a minimum cross-section of 30 mm × 5 mm will be needed at not more than 2.000 centres. If the floor does not have at least a 90 mm bearing on the supported wall lateral support should be provided by similar straps at not more than 2.000 centres or the joists should be fixed using restraint-type joist hangers.

PREVENTION OF DAMP IN CAVITY WALLS

Approved Document C recommends a cavity to be carried down at least 150 mm below the lowest damp-proof course and that any bridging of the cavity, other than a wall tie or closing course protected by the roof, is to have a suitable damp-proof course to prevent the passage of moisture across the cavity. Where the cavity is closed at the jambs of openings a vertical damp-proof course should be inserted unless some other suitable method is used to prevent the passage of moisture from the outer leaf to the inner leaf of the wall.

Approved Document C recommends a damp-proof course to be inserted in all external walls at least 150 mm above the highest adjoining ground or paving to prevent the passage of moisture rising up the wall and into the building, unless the design is such that the wall is protected or sheltered.

ADVANTAGES OF CAVITY WALL CONSTRUCTION

These can be listed as follows:

(a) Able to withstand a driving rain in all situations from penetrating to the inner wall surface.
(b) Gives good thermal insulation, keeping the building warm in winter and cool in the summer.
(c) No need for external rendering.
(d) Enables the use of cheaper and alternative materials for the inner construction.
(e) A nominal 255 mm cavity wall has a higher sound insulation value than a standard one brick thick wall.

DISADVANTAGES OF CAVITY WALL CONSTRUCTION

These can be listed as follows:

1. Requires a high standard of design and workmanship to produce a soundly constructed wall; this will require good supervision during construction.
2. The need to include a vertical damp-proof course to all openings.
3. Slightly dearer in cost than a standard one brick thick wall.

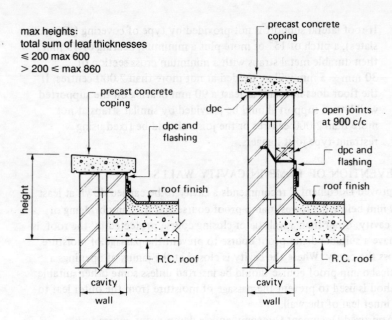

max heights:
total sum of leaf thicknesses
≤ 200 max 600
> 200 ≤ 860

precast concrete
coping

precast concrete
coping

dpc

open joints
at 900 c/c

dpc and
flashing

dpc and
flashing

roof finish

roof finish

height

R.C. roof

R.C. roof

cavity

cavity

wall

wall

Fig. II.23 Parapets

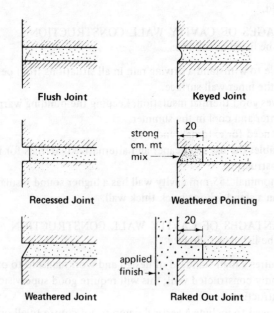

Flush Joint

Keyed Joint

strong
cm. mt
mix

20

Recessed Joint

Weathered Pointing

Weathered Joint

applied
finish

20

Raked Out Joint

Fig. II.24 Brick joints

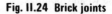

Jointing and pointing

These terms are used for the finish given to both the vertical and horizontal joints in brickwork irrespective of whether the wall is of brick, block, solid or cavity construction.

Jointing is the finish given to the joints when carried out as the work proceeds.

Pointing is the finish given to the joints by raking out to depth of approximately 20 mm and filling in on the face with a hard setting cement mortar which could have a colour additive. This process can be applied to both new and old buildings. Typical examples of jointing and pointing are shown in Fig. II.24.

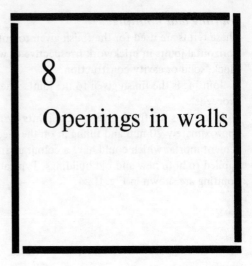

8
Openings in walls

An opening in an external wall consists of a head, jambs or reveals and a sill or threshold.

Head

The function of a head is to carry the triangular load of brickwork over the opening and transmit this load to the jambs at the sides. To fulfil this task it must have the capacity to support the load without unacceptable deflection. A variety of materials and methods is available in the form of a lintel or beam such as:

Timber: suitable for light loads and small spans, the timber should be treated with a preservative to prevent attack by beetles or fungi.

Steel:
For small openings—a mild steel flat or angle section can be used to carry the outside leaf of a cavity wall, the inner leaf being supported by a concrete or steel lintel.
For medium spans—a channel or joist section is usually suitable.
For large spans—a universal beam section to design calculations will be needed.
Steel lintels which are exposed to the elements should be either galvanised or painted with several coats of bituminous paint to give them protection against corrosion.

Concrete: these can be designed as *in situ* or precast reinforced beams or lintels and can be used for all spans. Prestressed concrete lintels are available for the small and medium spans.

Stone: these can be natural, artificial or reconstructed stone but are generally used as a facing to a steel or concrete lintel (see Fig. II.5).

Brick: unless reinforced with mild steel bars or mesh, brick lintels are only suitable for small spans up to 1 m, but like stone, bricks are also employed as a facing to a steel or concrete lintel.

Lintels require a bearing at each end of the opening, the amount will vary with the span but generally it will be 100 mm for the small spans and up to 225 mm for the medium and large spans. In cavity walling a damp-proof course will be required where the cavity is bridged by the lintel and this should extend at least 150 mm beyond each end of the lintel. Open joints are sometimes used to act as weep holes; these are placed at 900 mm centres in the outer leaf immediately above the damp-proof course. Typical examples of head treatments to openings are shown in Fig. II.25.

Arches

These are arrangements of wedged shaped bricks designed to support each other and carry the load over the opening round a curved profile to abutments on either side—full details of arch construction are given in the next chapter.

Jambs

In solid walls these are bonded to give the required profile and strength; examples of bonded jambs are shown in Fig. II.11. In cavity walls the cavity can be closed at the opening by using a suitable frame or by turning one of the leaves towards the other forming a butt joint in which is incorporated a vertical damp-proof course as required by the Building Regulations. Typical examples of jamb treatments to openings are shown in Fig. II.26.

Sill

The function of a sill is to shed the rain water, which has run down the face of the window or door and collected at the base, away from the opening and the face of the wall. Many methods and materials are available; appearance and durability are the main requirements since a sill is not a member which is needed to carry heavy loads. Sills, unlike lintels, do not require a bearing at each end. Typical examples of sill treatments to openings are shown in Fig. II.27.

71

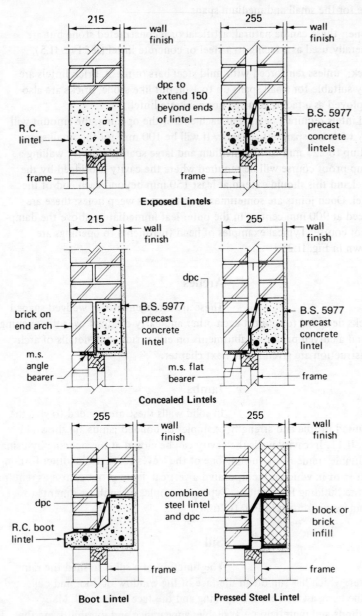

Fig. 11.25 Typical head treatments to openings

Labels in the figure:

Top left (Exposed Lintels): 215, wall finish, R.C. lintel, frame

Top right: 255, wall finish, dpc to extend 150 beyond ends of lintel, B.S. 5977 precast concrete lintels, frame

Exposed Lintels

Middle left (Concealed Lintels): 215, wall finish, brick on end arch, B.S. 5977 precast concrete lintel, m.s. angle bearer

Middle right: 255, wall finish, dpc, B.S. 5977 precast concrete lintel, m.s. flat bearer, frame

Concealed Lintels

Bottom left (Boot Lintel): 255, wall finish, dpc, R.C. boot lintel, frame

Bottom right (Pressed Steel Lintel): 255, wall finish, combined steel lintel and dpc, block or brick infill, frame

Boot Lintel   **Pressed Steel Lintel**

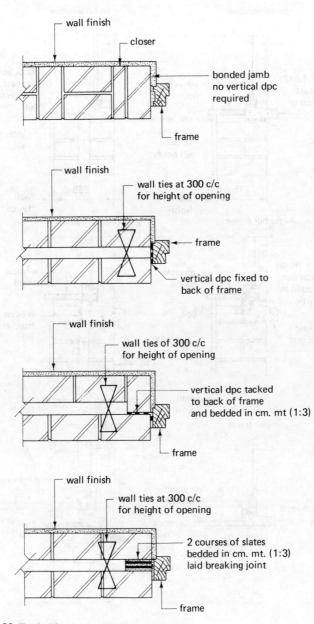

Fig. II.26 Typical jamb treatments to openings

73

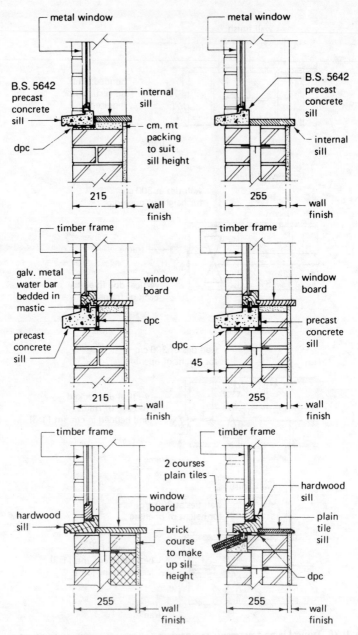

Fig. II.27 Typical sill treatments to openings

74

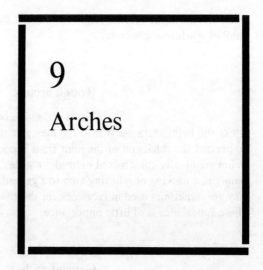

9
Arches

These are arrangements of wedge shaped bricks called 'voussoirs' which are designed to support each other and carry the load over the opening, round a curved profile, to abutments on either side. An exception to this form is the flat or 'soldier' arch constructed of bricks laid on end or on edge.

When constructing an arch it must be given temporary support until the brick joints have set and the arch has gained sufficient strength to support itself and carry the load over the opening. These temporary supports are called centres and are usually made of timber; their design is governed by the span, load and thickness of the arch to be constructed.

Soldier arches

This type of arch consists of a row of bricks showing on the face either the end or the edge of the bricks. Soldier arches have no real strength and if the span is over 1 000 mm they will require some form of permanent support such as a metal flat or angle (see Fig. II.25). If permanent support is not given the load will be transferred to the head of the frame in the opening instead of the jambs on either side. Small spans can have an arch of bonded brickwork by inserting into the horizontal joints immediately above the opening some form of reinforcement such as expanded metal or bricktor which is a woven strip of high tensile steel wires designed for the reinforcement of brick and stone walls. It is also possible to construct a soldier arch by inserting metal cramps in the vertical joints and casting these into an *in situ* backing lintel of reinforced concrete.

Rough arches

These arches are constructed of ordinary uncut bricks and being rectangular in shape they give rise to wedge shaped joints. To prevent the thick end of the joint from becoming too excessive rough arches are usually constructed in header courses. The rough arch is used mainly as a backing or relieving arch to a gauged brick or stone arch but they are sometimes used in facework for the cheaper form of building or where appearance is of little importance.

Gauged arches

These are the true arches and are constructed of bricks cut to the required wedge shape called voussoirs. The purpose of voussoirs is to produce a uniform thin joint which converges on to the centre point or points of the arch. There are two methods of cutting the bricks to the required wedge shape, namely, axed and rubbed. If the brick is of a hard nature it is first marked with a tin saw, to produce a sharp arris, and then it is axed to the required profile. For rubbed brick arches a soft brick called a rubber is used; the bricks are first cut to the approximate shape with a saw and are then finished off with an abrasive stone or file to produce the sharp arris. In both cases a template of plywood or hardboard to the required shape will be necessary for marking out the voussoirs. Typical examples of stone arches are shown in Fig. II.4; the terminology and setting out of simple brick arches is shown in Fig. II.28.

CENTRES

These are temporary structures, usually of light timber construction, which are strong enough to fulfil their function of supporting arches of brick or stone while they are being built and until they are sufficiently set to support themselves and the load over the opening. Centres can be an expensive item to a builder, therefore their design should be simple and adaptable so that as many uses as possible can be obtained from any one centre. A centre is always less in width than the soffit of an arch to allow for plumbing, that is, alignment and verticality of the face with a level or rule

76

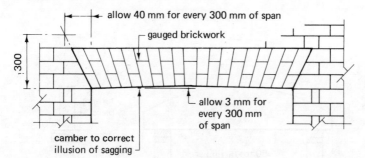

allow 40 mm for every 300 mm of span

gauged brickwork

300

allow 3 mm for every 300 mm of span

camber to correct illusion of sagging

Camber Arch

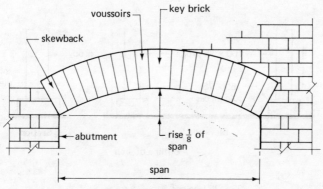

voussoirs

key brick

skewback

abutment

rise $\frac{1}{8}$ of span

span

Gauged Segmental Arch

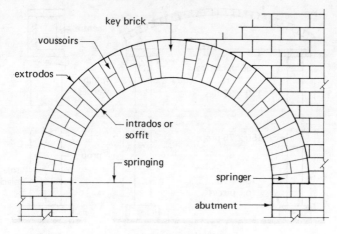

key brick

voussoirs

extrodos

intrados or soffit

springing

springer

abutment

Gauged Semi-circular Arch

Fig. II.28 Typical examples of brick arches

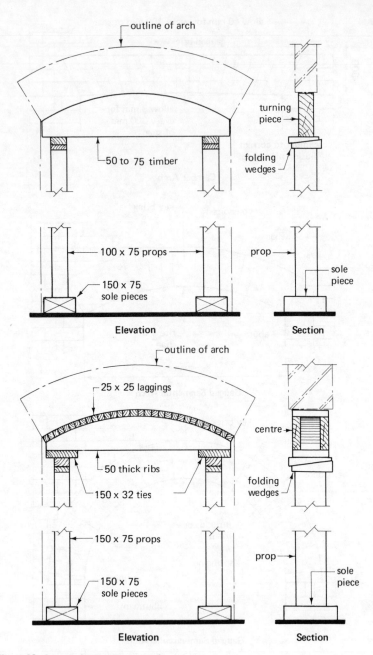

Fig. II.29 Centres for small span arches

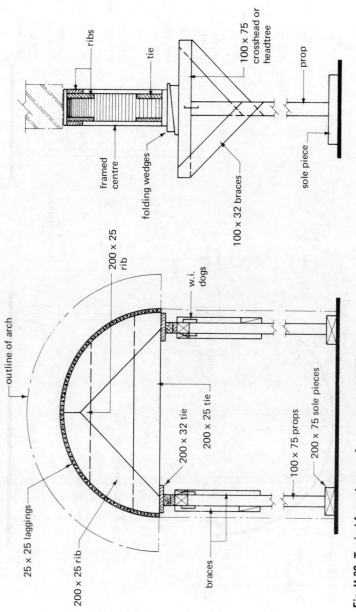

ribs

tie

100 × 75 crosshead or headtree

prop

framed centre

folding wedges

100 × 32 braces

sole piece

outline of arch

200 × 25 rib

w.i. dogs

25 × 25 laggings

200 × 25 rib

200 × 32 tie

200 × 25 tie

braces

100 × 75 props

200 × 75 sole pieces

Fig. II.30 Typical framed centre for spans up to 1500 mm

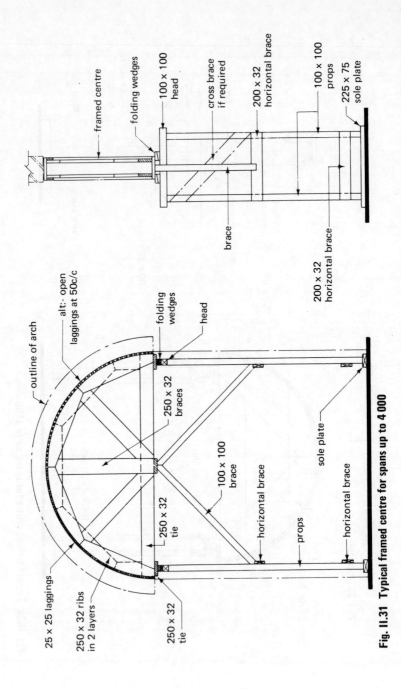

Fig. II.31 Typical framed centre for spans up to 4 000

framed centre

folding wedges

100 × 100 head

cross brace if required

200 × 32 horizontal brace

100 × 100 props

225 × 75 sole plate

brace

200 × 32 horizontal brace

outline of arch

alt:- open laggings at 50c/c

folding wedges

head

250 × 32 braces

100 × 100 brace

250 × 32 tie

250 × 32 tie

25 × 25 laggings

250 × 32 ribs in 2 layers

horizontal brace

props

horizontal brace

sole plate

The type of centre to be used will depend upon:

1. The weight to be supported.
2. The span.
3. The width of the soffit.

Generally soffits not wider than 150 mm will require one rib at least 50 mm wide and are usually called turning pieces. Soffits from 150-350 mm require two ribs which are framed together using horizontal tie members called laggings. Soffits over 350 mm require three or more sets of ribs. The laggings are used to tie the framed ribs together and to provide a base upon which the arch can be built. Close laggings are those which are touching each other, forming a complete seating for a gauged arch, and open laggings spaced at twice the width of the laggings, centre to centre, are used for rough arches.

If the arch is composed of different materials, for example, a stone arch with a relieving arch of brickwork, a separate centre for each material should be used. Typical examples of centres for brick arches are shown in Figs. II.29, II.30 and II.31.

10
Flooring

SOLID GROUND FLOORS

The construction of a solid ground floor can be considered under three headings:

1. Hardcore.
2. Blinding.
3. Concrete bed or slab.

Hardcore

The purpose of hardcore is to fill in any small pockets which have formed during oversite excavations, to provide a firm base on which to place a concrete bed and to help spread any point loads over a greater area. It also acts against capillary action of moisture within the soil. Hardcore is usually laid in 100-150 mm layers to the required depth and it is important that each layer is well compacted, using a roller if necessary, to prevent any unacceptable settlement beneath the solid floor.

Approved Document C recommends that no hardcore laid under a solid ground floor shall contain water-soluble sulphates or other harmful matter in such quantities as to be liable to cause damage to any part of the floor. This recommendation prevents the use of any material which may swell upon becoming moist, such as colliery shale, and furthermore it is necessary to ascertain that brick rubble from demolition works and clinker furnace waste intended for use as hardcore does not have any harmful water-soluble sulphate content.

Blinding

This is used to even off the surface of hardcore if a damp-proof membrane is to be placed under the concrete bed or if a reinforced concrete bed is specified. Firstly, it will prevent the damp-proof membrane from being punctured by the hardcore and, secondly, it will provide a true surface from which the reinforcement can be positioned. Blinding generally consists of a layer of fine ash or sand 25-50 mm thick or a 50-75 mm layer of weak concrete (1 : 12 mix usually suitable) if a true surface for reinforced concrete is required.

Concrete bed

Thicknesses generally specified are:

1. Unreinforced or plain *in situ* concrete, 100-150 mm thick.
2. Reinforced concrete, 150 mm minimum.

Suitable concrete mixes are:

(*a*) Plain *in situ* concrete, 1 : 3 : 6 or 1 : 6 'all-in'.
(*b*) Reinforced concrete, 1 : 2 : 4.

The reinforcement used in concrete beds for domestic work is usually in the form of a welded steel fabric to BS 4483. Sometimes a light square mesh fabric is placed 25 mm from the upper surface of the concrete bed to prevent surface crazing and limit the size of any cracking.

In domestic work the areas of concrete are defined by the room sizes and it is not usually necessary to include expansion or contraction joints in the construction of the bed.

PROTECTION OF FLOORS NEXT TO THE GROUND

Building Regulation C4 requires that such part of a building as is next to the ground shall have a floor so constructed as to prevent the passage of moisture from the ground to the upper surface of the floor. The requirements of this regulation can only be properly satisfied by the provision of a suitable barrier in the form of a damp-proof membrane within the floor. The membrane should be turned up at the edges to meet and blend with the damp-proof course in the walls to prevent any penetration of moisture by capillary action at edges of the bed.

Suitable materials for damp-proof membranes are:

1. Waterproof building papers—conforming to BS 1521.
2. Polythene sheet 1 000 gauge sheet with sealed joints is acceptable and will also give protection against moisture vapour as well as moisture.
3. Hot poured bitumen—should be at least 3 mm thick.
4. Cold applied bitumen/rubber emulsions—should be applied in not less than three coats.
5. Asphalt—could be dual purpose finish and damp-proof membrane.

The position of a damp-proof membrane, whether above or below the concrete bed, is a matter of individual choice. A membrane placed above the bed is the easiest method from a practical aspect and is therefore generally used. A membrane placed below the bed has two advantages: firstly, it will keep the concrete bed dry and in so doing will make the bed a better thermal insulator and, secondly, during construction it will act as a separating layer preventing leakage of the cement matrix into the hardcore layer which could result in a weak concrete mix. Typical details of solid floor construction are shown in Figs. II.32 and II.33.

SUSPENDED TIMBER GROUND FLOORS

This type of floor consists of timber boards or other suitable sheet material fixed to joists spanning over sleeper walls and was until 1939 the common method of forming ground floors in domestic buildings. The Second World War restricted the availability of suitable timber and solid ground floors replaced suspended timber floors. Today the timber floor is still used on occasions because it has some flexibility and will easily accept nail fixings—properties which a solid ground floor lacks. It is a more expensive form of construction than a solid floor and can only be justified on sloping sites which would need a great deal of filling to make up the ground to the required floor level.

Suspended timber ground floors are susceptible to dry rot, draughts and are said to be colder than other forms of flooring. If the floor is correctly designed and constructed these faults can be eliminated.

The problem of dry rot, which is a fungus that attacks damp timber, can be overcome by adequate ventilation under the floor and the correct positioning of damp-proof courses to keep the under floor area and timber dry. Through ventilation is essential to keep the moisture content of the timber below that which would allow fungal growth to take place; that is, 20% of its oven-dry weight. The usual method is to allow a free flow of air under the floor covering by providing, in the external walls, air bricks sited near the corners and at approximately 2 000 mm centres

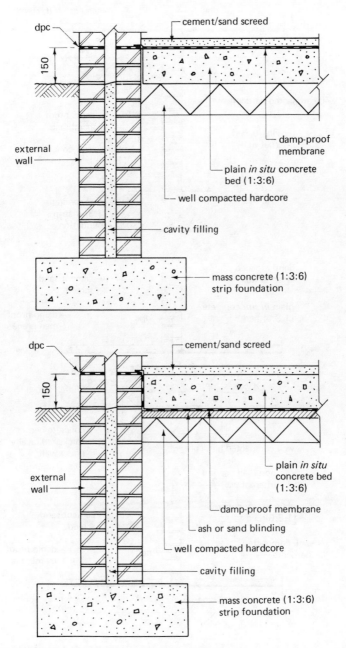

Fig. II.32 Typical solid floor details at external walls

85

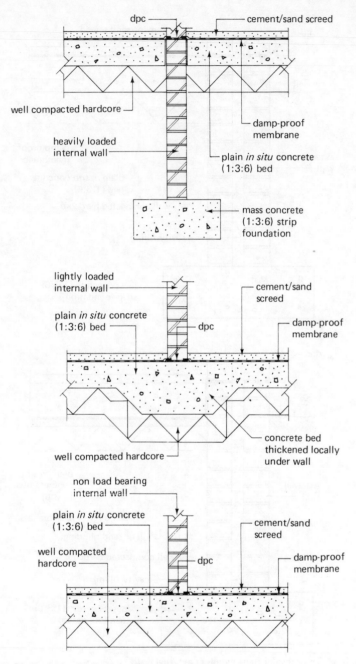

Fig. II.33 Typical solid floor details at internal walls

around the perimeter of the building. If a suspended timber floor is used in conjunction with a solid ground floor in an adjoining room, pipes are used under the solid floor to convey air to and from the external walls.

BUILDING REGULATIONS

Building Regulation C4 applies as with solid floors and recommended provisions are given in Approved Document C. Figure II.34 shows the minimum dimensions recommended in Approved Document C but in practice a greater space between the concrete bed and the timber is usual. The honeycomb sleeper walls are usually built two or three courses high to allow good through ventilation. Sleeper walls spaced at 2 000 mm centres will give an economic joist size. The width of joists is usually taken as 50 mm, this will give sufficient width for the nails securing the covering, and the depth can be obtained by reference to Table B3 in Approved Document A or by design calculations. The usual joist depth for domestic work is 125 mm.

Lay-out

The most economic lay-out is to span the joists across the shortest distance of the room, this means that joists could be either parallel or at right-angles to a fireplace. The fireplace must be constructed of non-combustible materials and comply with Building Regulation J3. Typical examples are shown in Figs. II.35 and II.36.

SUSPENDED TIMBER UPPER FLOORS

Timber, being a combustible material, is restricted by Part B of the Building Regulations to small domestic buildings as a structural flooring material. Its popularity in this context, is due to its low cost in relationship to other structural flooring methods and materials. Structural softwood is readily available at a reasonable cost, is easily worked and has a good strength to weight ratio and is therefore suitable for domestic loadings.

Terminology

Common joist: a joist spanning from support to support.

Trimming joist: span as for common joist but it is usually 25 mm thicker and supports a trimmer joist.

Trimmer joist: a joist at right-angles to the main span supporting the trimmed joists and is usually 25 mm thicker than a common joist.

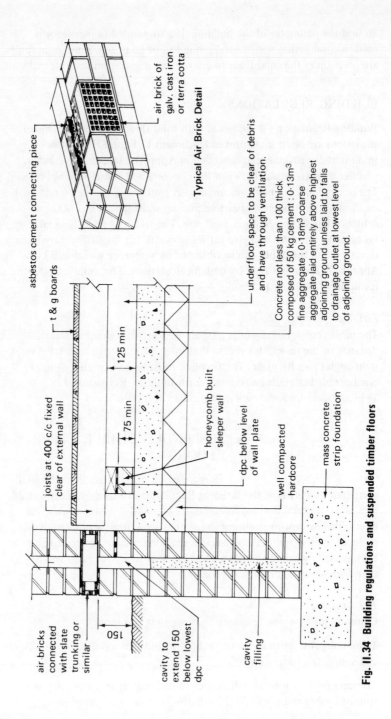

air bricks connected with slate trunking or similar

150

cavity to extend 150 below lowest dpc

cavity filling

mass concrete strip foundation

t & g boards

joists at 400 c/c fixed clear of external wall

125 min

75 min

honeycomb built sleeper wall

dpc below level of wall plate

well compacted hardcore

underfloor space to be clear of debris and have through ventilation.

Concrete not less than 100 thick composed of 50 kg cement : 0·13m³ fine aggregate : 0·18m³ coarse aggregate laid entirely above highest adjoining ground unless laid to falls to drainage outlet at lowest level of adjoining ground.

air brick of galv. cast iron or terra cotta

asbestos cement connecting piece

Typical Air Brick Detail

Fig. II.34 Building regulations and suspended timber floors

88

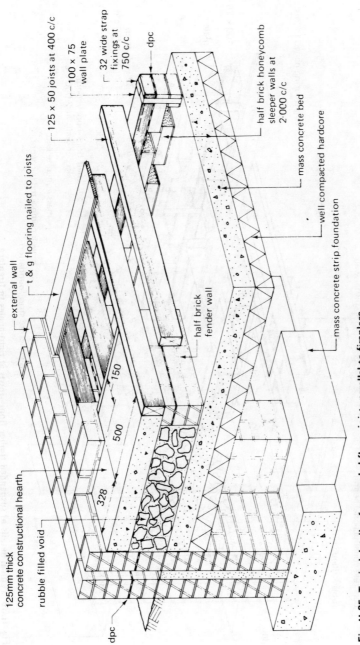

125mm thick concrete constructional hearth

rubble filled void

external wall

t & g flooring nailed to joists

125 × 50 joists at 400 c/c

100 × 75 wall plate

32 wide strap fixings at 750 c/c

dpc

half brick honeycomb sleeper walls at 2·000 c/c

mass concrete bed

well compacted hardcore

half brick fender wall

mass concrete strip foundation

150

500

328

dpc

Fig. II.35 Typical details of suspended floor—joists parallel to fireplace

89

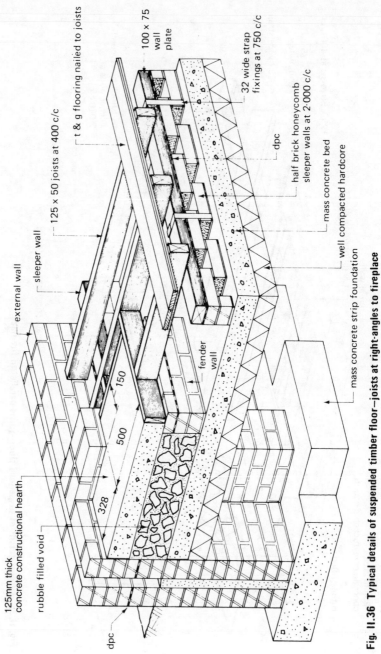

100 × 75 wall plate

t & g flooring nailed to joists

125 × 50 joists at 400 c/c

sleeper wall

external wall

150

500

328

125mm thick concrete constructional hearth

rubble filled void

dpc

fender wall

32 wide strap fixings at 750 c/c

dpc

half brick honeycomb sleeper walls at 2.000 c/c

mass concrete bed

well compacted hardcore

mass concrete strip foundation

Fig. II.36 Typical details of suspended timber floor—joists at right-angles to fireplace

Trimmed joist: a joist cut short to form an opening and is supported by a trimmer joist; it spans in the same direction as common joists and is of the same section size.

Joist Sizing

There are three ways of selecting a suitable joist size for supporting a domestic type floor:

1. Rule of thumb: $\dfrac{\text{span in mm}}{24} + 50$ mm = depth in mm.

2. Calculation: $BM = \dfrac{fbd^2}{6}$ where BM = bending moment

f = maximum fibre stress
b = breadth (assumed to be 50 mm)
d = depth in mm.

3. Approved Document A, Tables B3 and B4.

JOISTS

If the floor is framed with structural softwood joists of a size not less than that required by the Approved Document, the usual width is taken as 50 mm. The joists are spaced at 375-450 mm centre to centre depending on the width of the ceiling boards which are to be fixed on the underside. Maximum economy of joist size is obtained by spanning in the direction of the shortest distance to keep within the deflection limitations allowed. The maximum economic span for joists is between 3 500 and 4 500 mm, for spans over this a double floor could be used.

Support

The ends of the joists must be supported by load bearing walls. The common methods are to build in the ends or to use special metal fixings called joist hangers; other methods are possible but these are seldom employed. Support on internal load bearing walls can be by joist hangers or direct bearing when the joists are generally lapped (see Fig. II.37).

Trimming

This is a term used to describe the framing of joists around an opening or projection. Various joints can be used to connect the members together, all of which can be substituted by joist hangers. Trimming around flues and upper floor fireplaces should comply with the recommendations of Approved Document J. It should be noted that, since central heating is becoming commonplace, the provision of upper floor fireplaces is seldom

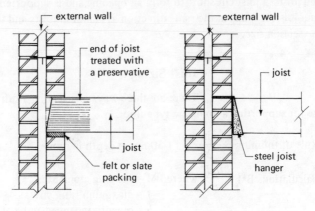

Wall Bearing

external wall

end of joist treated with a preservative

joist

felt or slate packing

Joist Hanger Bearing

external wall

joist

steel joist hanger

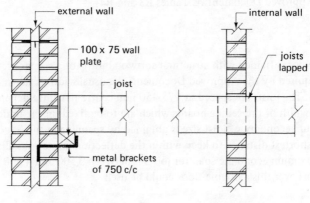

Plate Bearing

external wall

100 x 75 wall plate

joist

metal brackets of 750 c/c

Direct Bearing

internal wall

joists lapped

fishtail for building into brickwork

straps for fixing to timber

Typical Joist Hangers

Fig. II.37 Typical joist support details

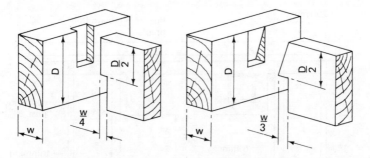

Housed Joint

Bevelled Housed Joint

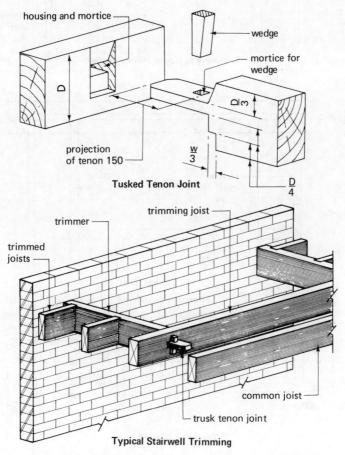

housing and mortice

wedge

mortice for wedge

D

projection of tenon 150

$\frac{w}{3}$

$\frac{D}{3}$

$\frac{D}{4}$

Tusked Tenon Joint

trimming joist

trimmer

trimmed joists

common joist

trusk tenon joint

Typical Stairwell Trimming

Fig. II.38 Floor trimming joints and details

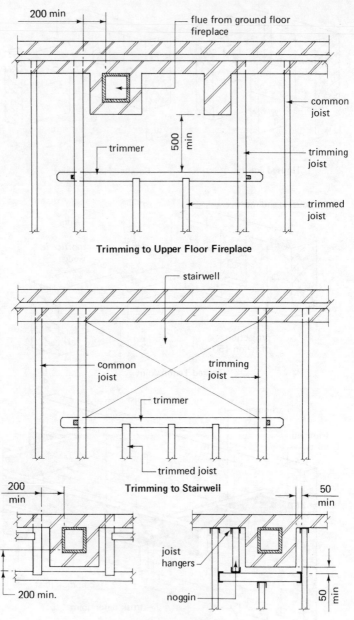

Trimming to Upper Floor Fireplace

Trimming to Stairwell

Trimming Around Flues

Fig. II.39 Typical trimming arrangements

94

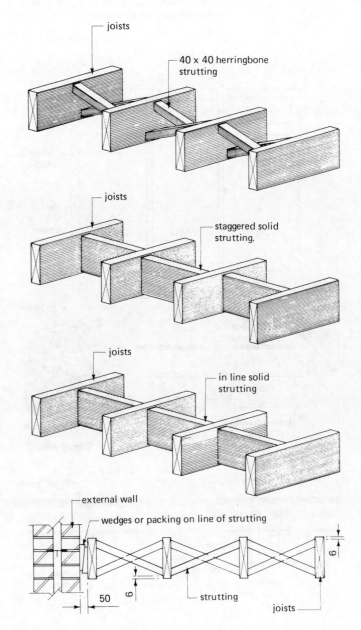

Fig. II.40 Strutting arrangements

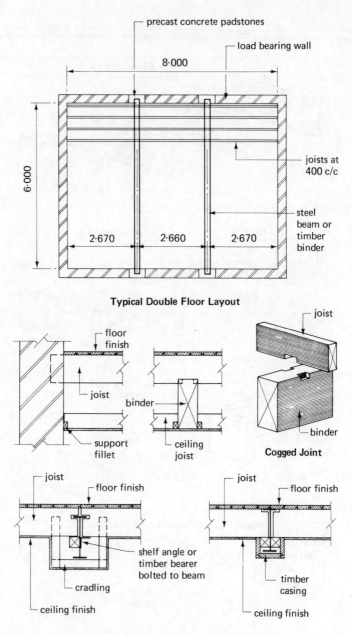

Typical Double Floor Layout

precast concrete padstones

load bearing wall

8·000

6·000

2·670   2·660   2·670

joists at 400 c/c

steel beam or timber binder

floor finish

joist

support fillet

binder

ceiling joist

joist

binder

Cogged Joint

joist

floor finish

shelf angle or timber bearer bolted to beam

cradling

ceiling finish

joist

floor finish

timber casing

ceiling finish

Typical Details Using Timber Binder or Steel Beam

Fig. II.41 Double floor details

included in modern designs, because they are considered to be superfluous. Typical trimming joints and arrangements are shown in Figs. II.38 and II.39.

Strutting

Shrinkage in timber joists will cause twisting to occur and this will result in movement of the ceiling below and could cause the finishes to crack. To prevent this strutting is used between the joists if the total span exceeds 2 400 mm; the strutting being placed at mid-span (see Fig. II.40).

DOUBLE FLOORS

These can be used on spans over 4 500 mm to give a lower floor area free of internal walls. They consist of a steel beam or timber binder spanning the shortest distance which supports common joists spanning at right-angles. The beam reduces the span of the common joists to a distance which is less than the shortest span to allow an economic joist section to be used. The use of a timber binder was a popular method but it is generally considered to be uneconomic when compared with a standard steel beam section. Typical details are shown in Fig. II.41.

If the span is such that a double floor is deemed necessary it would be a useful exercise to compare the cost with that of other flooring methods, such as *in situ* reinforced concrete and precast concrete systems, which, overall, could be a cheaper and more practical solution to the problem.

11
Roofs–timber, flat and pitched

TIMBER FLAT ROOFS

A flat roof is essentially a low pitched roof and is defined in BS 3589 as a pitch of $10°$ or less to the horizontal. Generally the angle of pitch is governed by the type of finish which is to be applied to the roof.

The functions of any roof are:

1. To keep out rain, wind, snow and dust.
2. To prevent excessive heat loss in winter.
3. To keep the interior of the building cool in summer.
4. Designed to accommodate all stresses encountered.
5. Designed to accept movement due to changes in temperature and moisture content.

The simplest form of roof construction to fulfil these functions is a timber flat roof covered with an impervious material to prevent rain penetration. This form of roof is suitable for spans up to 4 000 mm, spans over this are usually covered with a reinforced concrete slab or a patent form of decking.

The disadvantages of timber flat roofs are:

(a) They are poor insulators against the transfer of heat.
(b) They tend to give low rise buildings an unfinished appearance.
(c) Unless properly designed and constructed pools of water will collect on the surface causing local variations in temperature. This

results in deterioration of the covering and, consequently, high maintenance costs.

Construction

The construction of a timber flat roof follows the same methods as those employed for the construction of timber upper floors. Suitable joist sizes can be obtained by design or by reference to Tables B21 and B22 given in Approved Document A. The spacing of roof joists is controlled by the width of decking material to be used and/or the width of ceiling board on the underside. Timber flat roofs are usually constructed to fall in one direction towards a gutter or outlet. This can be achieved by sloping the joists to the required fall but, as this would give a sloping soffit on the underside, it is usual to fix wedge shaped fillets called 'firrings' to the top of the joists to provide the fall (see Fig. II.42). The materials used in timber flat roof construction are generally poor thermal insulators and therefore some form of non-structural material can be incorporated into the roof if it has to comply with Part L of the Building Regulations.

DECKING MATERIALS

Timber: This can be in the form of softwood boarding, chipboard or plywood. Plain edge sawn boards or tongued and grooved boards are suitable for joists spaced at centres up to 450 mm. Birch or fir plywood is available in sheet form which requires to be fixed on all four edges—this means that noggins will be required between joists to provide the bearing for the end fixings. Chipboard is also a sheet material and is fixed in a similar manner to plywood by using noggins and it should be noted that this material is susceptible to moisture movement. Flat roofs using a timber decking should have the roof void ventilated to minimise moisture content fluctuations; therefore it is advisable to use structural timbers which have been treated against fungal and insect attack. In certain areas treatment to prevent infestation by the house longhorn beetle is required under Building Regulation 7 (Table 1 in supporting AD).

Compressed straw slabs: these are made from selected straw by a patent method of heat and pressure to a standard width of 1 200 mm with a selection of lengths, the standard thickness being 50 mm which gives sufficient strength for the slabs to span 600 mm. All edges of the slabs must be supported and fixed. Ventilation is of the utmost importance and it is common practice to fix cross bearers at right-angles to and over the joists to give cross ventilation. A bitumen scrim should be placed over the joints before the weathering membrane is applied.

Wood wool slabs: these are 600 mm wide slabs of various lengths and thicknesses which can span up to 1 200 mm. The slabs are made of shredded wood fibres which have been chemically treated and are bound together with cement. The fixing and laying is similar to compressed straw slabs.

INSULATING MATERIALS

There are many types of insulating materials available, usually in the form of boards or quilts. Boards are laid over the joists, either under or on top of the rough boarding, whereas quilted materials are laid over or between the joists.

Boards: these are made from lightly compressed vegetable fibres which are bonded together with glues or resins. Being lightly compressed they contain a large number of tiny voids in which air is trapped and it is this entrapped air which makes them good thermal insulators. Since these boards are light in weight, low in strength and are compressible they must have adequate support and fixings. Fibreboards are made to a wide variety of sizes, the thicker boards being the better insulators.

Compressed straw slabs and wood wool slabs are also good thermal insulators and can be used in the dual role of decking and insulation.

Quilts: these are made from mineral or glass wool which is loosely packed between sheets of paper, since they are fine shreds giving rise to irritating scratches if handled. Quilts rely on the loose way in which the core is packed for their effectiveness and therefore the best results are obtained when they are laid between joists.

A variety of loose fills are also available for placing between the joists and over the ceiling to act as thermal insulators.

WEATHER-PROOF FINISHES

Suitable materials are asphalt, lead, copper, zinc, aluminium and built-up roofing felt; only the latter will be considered at this stage.

Built-up roofing felt

Most roofing felts consist essentially of a base sheet which is impregnated with hot bitumen during manufacture and is then coated on both sides with a stabilised weatherproof bitumen compound. This outer coating is dusted while still hot and tacky with powdered talc or receives an application of fine or coarse sand or coloured mineral granules. After

100

cooling the felt is cut to form rolls 1 m wide and 10 or 20 m long before being wrapped for dispatch.

BS 747 Part 2 specifies three classes of base felt:

Class 1: fibre base—this type is very flexible and of low cost.
Class 2: asbestos base—this type has good fire resistance, is fairly stable and is virtually rot proof.
Class 3: glass fibre base—this type is rot proof, very stable and is used for high quality work.

For flat roofs three layers of felt should be used the first being laid at right-angles to the fall commencing at the eaves. If the decking is timber the first layer is secured with large flat-head felt nails and the subsequent layers are bonded to it with a hot bitumen compound by a roll and pour method. If the decking is of a material other than timber all three layers are bonded with hot bitumen compound. It is usually recommended that a vented first layer is used in case moisture is trapped during construction; this recommendation does not normally apply to roofs with a timber deck since timber has the ability to 'breathe'. The minimum fall recommended for built-up roofing felt is 17 mm in 1 000 mm or 1°.

In general the Building Regulations require a flat roof with a weather-proofing which has a surface finish of asbestos based bituminous felt or is covered with a layer of stone chippings. The chippings protect the under-lying felt, provide additional fire resistance and give increased solar reflection. A typical application would be 12·5 mm stone chippings at approximately 50 kilogrammes to each 2·5 square metres of roof area. Chippings of limestone, granite and light coloured gravel would be suitable.

VAPOUR BARRIERS

The problem of condensation should always be considered when con-structing a flat roof. The insulation below the built-up roofing felt will not prevent condensation occurring and since it is a permeable material water vapour will pass upwards through it and condense on the underside of the roofing felt. The drops of moisture so formed will soak into the insulating material lowering its insulation value and possibly causing staining on the underside. To prevent this occurring a vapour barrier should be placed on the underside of the insulating material. A layer of roofing felt with sealed laps is suitable, an alternative being a layer of aluminium foil (many boards can be obtained with the foil already attached).

For typical timber flat roof details see Fig. II.42.

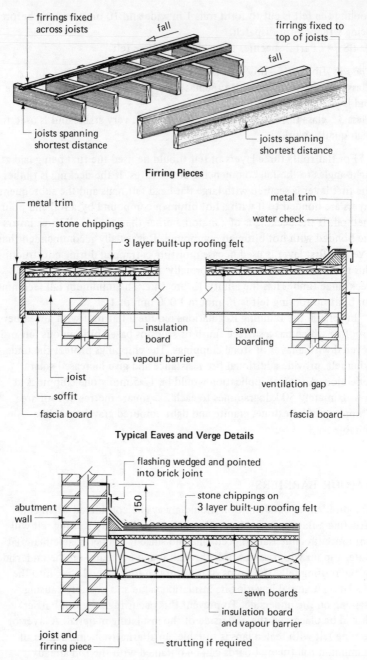

Firring Pieces

- firrings fixed across joists
- fall
- firrings fixed to top of joists
- fall
- joists spanning shortest distance
- joists spanning shortest distance

- metal trim
- stone chippings
- 3 layer built-up roofing felt
- metal trim
- water check
- insulation board
- sawn boarding
- vapour barrier
- joist
- ventilation gap
- soffit
- fascia board
- fascia board

Typical Eaves and Verge Details

- flashing wedged and pointed into brick joint
- stone chippings on 3 layer built-up roofing felt
- 150
- abutment wall
- sawn boards
- insulation board and vapour barrier
- joist and firring piece
- strutting if required

Fig. II.42 Timber flat roof details

TIMBER PITCHED ROOFS

The term pitched roof includes any roof whose angle of slope to the horizontal lies between 10° and 70°; below this range it would be called a flat roof and above 70° it would be classified as a wall.

The pitch is generally determined by the covering which is to be placed over the timber carcass, whereas the basic form is governed by the load and span. The terminology used in timber roof work and the basic members for various spans are shown in Figs. II.43 and II.44.

The cost of constructing a pitched roof is comparable to that of a flat roof but a pitched roof also provides a useful void in which to house a cold water storage cistern. The timber used in roof work is structural softwood, the members being joined together with nails. The sloping components or rafters are used to transfer the covering, wind, rain and snow loads to the load-bearing walls on which they rest. The rafters are sometimes assisted in this function by struts and purlins in what is called a purlin or double roof (see Fig. II.45). As with other forms of roofs the spacing of the rafters and consequently the ceiling joists is determined by the module size of the ceiling boards which are to be fixed on the underside of the joists.

Roof members

Ridge: this is the spine of a roof and is essentially a pitching plate for the rafters which are nailed to each other through the ridge board. The depth of ridge board is governed by the pitch of the roof, the steeper the pitch the deeper will be the vertical or plumb cuts on the rafters abutting the ridge.

Common rafters: the main load-bearing members of a roof, they span between a wall plate at eaves level and the ridge. Rafters have a tendency to thrust out the walls on which they rest and this must be resisted by the walls and the ceiling joists. Rafters are notched over and nailed to a wall plate situated on top of a load-bearing wall, the depth of the notch should not exceed one-third the depth of the rafter.

Jack rafters: these fulfil the same function as common rafters but span from ridge to valley rafter or from hip rafter to wall plate.

Hip rafters: similar to a ridge but forming the spine of an external angle and similar to a rafter spanning from ridge to wall plate.

Valley rafters: as hip rafter but forming an internal angle.

Wall plates: these provide the bearing and fixing medium for the various roof members and distribute the loads evenly over the supporting walls; they are bedded in cement mortar on top of the load-bearing walls.

Dragon ties: a tie place across the corners and over the wall plates to help provide resistance to the thrust of a hip rafter.

Ceiling joists: these fulfil the dual function of acting as ties to the feet of pairs of rafters and providing support for the ceiling boards on the underside and any cisterns housed within the roof void.

Purlins: these act as beams reducing the span of the rafters enabling an economic section to be used. If the roof has a gable end they can be supported on a corbel or built in but in a hipped roof they are mitred at the corners and act as a ring beam.

Struts: these are compression members which transfer the load of a purlin to a suitable load-bearing support within the span of the roof.

Collars: these are extra ties to give additional strength and are placed at purlin level.

Binders: these are beams used to give support to ceiling joists and counteract excessive deflections and are used if the span of the ceiling joist exceeds 2 400 mm.

Hangers: vertical members used to give support to the binders and allow an economic section to be used, they are included in the design if the span of the binder exceeds 3 600 mm.

The arrangement of struts, collars and hangers only occurs on every fourth or fifth pair of rafters.

EAVES

The eaves of a roof is the lowest edge which overhangs the wall thus giving the wall a degree of protection, it also provides the fixing medium for the rainwater gutter. The amount of projection from the wall of the eaves is a matter of choice but is generally in the region of 300-450 mm.

Two basic types of eaves are used, open eaves and closed eaves. The former is a cheap method, the rafters being left exposed on the underside and should be treated with a preservative. The space between the rafters and the roof covering is filled in with brickwork, a process called beam filling and a continuous triangular fillet is fixed over the backs of the rafters to provide the support for the bottom course of slates or tiles. The closed eaves is where the feet of the rafters are boxed in using a vertical board called a 'fascia' and the space between the fascia and the wall being filled in with a soffit board, the brick wall being terminated above the soffit level (see Fig. II.45).

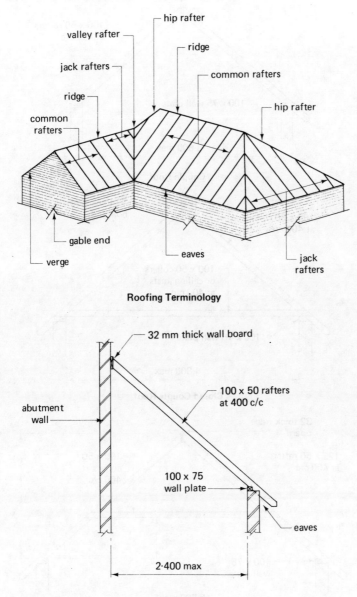

Roofing Terminology

Lean-to Roof

Fig. II.43 Roofing terminology and lean-to-roof

105

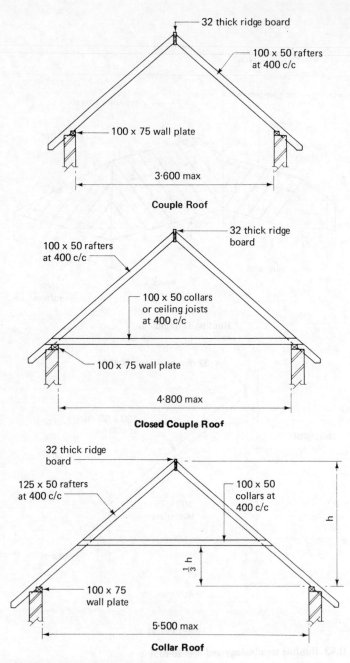

Fig. II.44 Pitched roofs for small spans

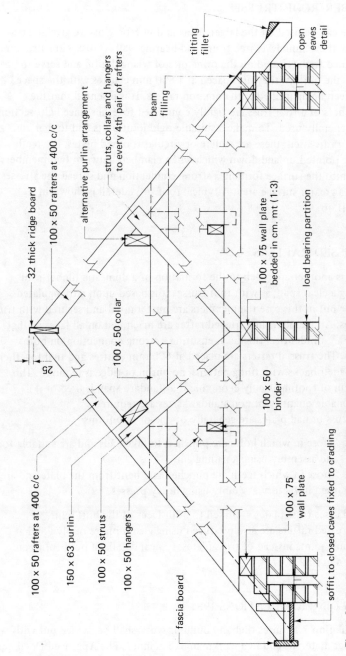

Fig. II.45 Typical double or purlin roof details for spans up to 7 200 mm

tilting fillet

open eaves detail

beam filling

32 thick ridge board

100 × 50 rafters at 400 c/c

alternative purlin arrangement

struts, collars and hangers to every 4th pair of rafters

100 × 50 collar

25

100 × 75 wall plate bedded in cm. mt (1:3)

load bearing partition

100 × 50 binder

100 × 50 rafters at 400 c/c

150 × 63 purlin

100 × 50 struts

100 × 50 hangers

100 × 75 wall plate

fascia board

soffit to closed eaves fixed to cradling

107

TIMBER ROOF TRUSSES

These can be used on the larger spans in domestic work to give an area below the ceiling level free from load-bearing walls. Trusses are structurally designed frames based on the principles of triangulation and serve to carry the purlins, they are spaced at 1 800 mm centres with the space between being infilled with common rafters. It is essential that the members of a roof truss are rigidly connected together since light sections are generally used. To make a suitable rigid joint bolts and timber connectors are used; these are square or circular toothed plates, the teeth being pointed up and down which when clamped between two members bite into the surface forming a strong connection and spread the stresses over a greater surface area. A typical roof truss detail is shown in Fig. II.46.

TRUSSED RAFTERS

This is another approach to the formation of a domestic timber roof giving a clear span; as with roof trusses it is based upon a triangulated frame but in this case the members are butt jointed and secured with truss plates. All members in a trussed rafter are machined on all faces so that they are of identical thickness ensuring a strong connection on both faces. The trussed rafters are placed at 600 mm centres and tied together over their backs with tiling battens, no purlin or ridge is required. This system of roofing is only economic if a standard span is used or if a reasonable quantity of non-standard sizes are required.

Truss or nail plates are generally of one or two forms:

1. Those in which holes are punched to take nails and are suitable for site assembly using a nailing gun.
2. Those in which teeth are punched and bent from the plate and are used in factory assembly using heavy presses.

In all cases truss plates are fixed to both faces of the butt joint.

Trussed rafters are also produced using gusset plates of plywood at the butt joints instead of truss plates, typical details of both forms are shown in Fig. II.47.

BUILDING REGULATIONS 1985

Regulation 7 requires that any building work shall be carried out with proper materials and in a workman-like manner. The Approved Document supporting this regulation defines acceptable levels of performance for materials which include products, fittings, items of equipment and back-

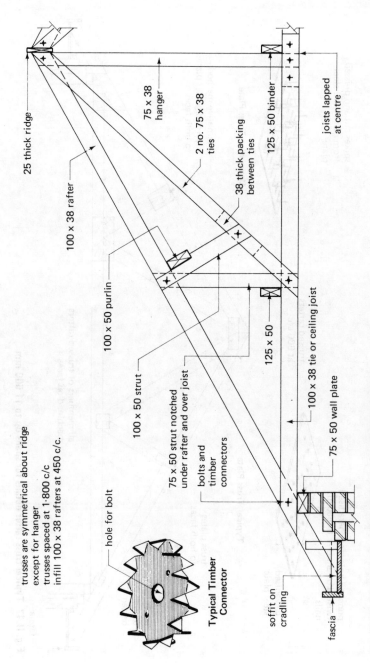

trusses are symmetrical about ridge
except for hanger
trusses spaced at 1·800 c/c
infill 100 × 38 rafters at 450 c/c.

25 thick ridge

75 × 38 hanger

100 × 38 rafter

2 no. 75 × 38 ties

38 thick packing between ties

125 × 50 binder

joists lapped at centre

100 × 50 purlin

100 × 50 strut

125 × 50

100 × 38 tie or ceiling joist

75 × 50 strut notched under rafter and over joist

bolts and timber connectors

75 × 50 wall plate

hole for bolt

Typical Timber Connector

soffit on cradling

fascia

Fig. II.46 Typical truss detail for spans up to 8 000 mm

109

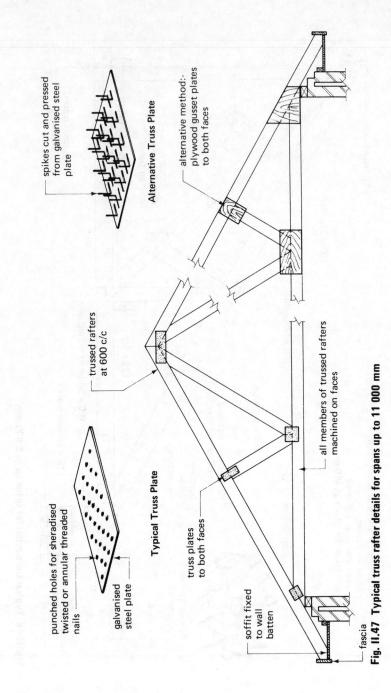

spikes cut and pressed from galvanised steel plate

Alternative Truss Plate

alternative method:- plywood gusset plates to both faces

trussed rafters at 600 c/c

all members of trussed rafters machined on faces

punched holes for sheradised twisted or annular threaded nails

galvanised steel plate

Typical Truss Plate

truss plates to both faces

soffit fixed to wall batten

fascia

Fig. II.47 Typical truss rafter details for spans up to 11 000 mm

filling for excavations. The aids for establishing fitness of materials suggested in the Approved Document are:

1. Past experience—such as a building in use.
2. Agrément Certificates.
3. British Standards.
4. Independent certification schemes.
5. Quality assurance schemes complying with BS 5750.
6. Test and calculations for materials only using the NATLAS accreditation scheme for testing laboratories.

The Approved Document supporting regulation 7 lists certain geographical areas where the softwood timber used for roof construction should be adequately treated with a suitable preservative to prevent infestation by the House Longhorn Beetle. This is a wise precaution whether it is recommended or not.

Regulation B4(2) requires the roof to offer adequate resistance to the spread of fire over the roof. Table 1.3 in Approved Document B gives limitations on roof coverings for dwelling houses by designations and distance from the boundary, the designations being defined in Table A4.

Regulation C4 requires that the roof of a building shall adequately resist the passage of moisture to the inside of the building.

Regulation L2 states the maximum thermal coefficients or 'U' values permitted in dwellings. It should be noted that the lower the numerical value of the coefficient the better is the insulation. For the roofs of dwellings the maximum 'U' value is 0.3 W/m² K. This value is not normally achieved with traditional roof constructions and therefore a suitable insulating material must be added.

The selection of suitable structural timber members for pitched roofs for single family houses of not more than three storeys can be ascertained by reference to Tables B5 to B20 in Approved Document A.

12

Roof tiling and slating

A roof can be defined as the upper covering of a building and in this position it is fully exposed to the rain, snow, wind, sun and general atmosphere, therefore the covering to any roof structure must have good durability to meet these conditions. Other factors to be taken into account are weight, maintenance and cost.

Roofs are subjected to wind pressures, both positive and negative, the latter causing uplift and suction which can be overcome by firmly anchoring lightweight coverings to the structure or by relying upon the deadweight of the covering material. Domestic roofs are not usually designed for general access and therefore the chosen covering material should be either very durable or easily replaceable. The total dead load of the covering will affect the type of support structure required and ultimately the total load on the foundations, so therefore careful consideration must be given to the medium selected for the roof covering.

In domestic work the roof covering usually takes the form of tiling or slating since these materials economically fulfil the above requirements and have withstood the test of time.

Tiling

Tiles are manufactured from clay and concrete to a wide range of designs and colours suitable for pitches from 17-45° and work upon the principle of either double or single lap. The vital factor for the efficient performance of any tile or slate is the pitch and it should be noted that the pitch of a tile is always less than the pitch of the rafters owing to the overlapping technique.

Tiles are laid in overlapping courses and rely upon the water being shed off the surface of one tile onto the exposed surface of the tile in the next course. The problem of water entering by capillary action between the tiles is overcome by the camber of the tile, the method of laying or by overlapping side joints. In all methods of tiling a wide range of fittings are produced to enable all roof shapes to be adequately protected.

PLAIN TILING

This is a common method in this country and works on the double lap principle. The tiles can be of hand-made or machine pressed clay, the process of manufacture being similar to that of brickmaking. The hand-made tile is used mainly where a rustic or distinctive roof character is required since they have a wide variation in colour, texture and shape. They should not be laid on a roof of less than $45°$ pitch since they tend to absorb water and if allowed to become saturated they may freeze, expand and spall or fracture in cold weather. Machine pressed tiles are harder, denser and more uniform in shape than hand-made varieties and can be laid to a minimum pitch of $35°$.

A suitable substitute for plain clay tiles is the concrete plain tile, these are produced in a range of colours to the same size specifications as the clay tiles and with the same range of fittings. The main advantage of concrete tiles is the lower cost and the main disadvantage is the extra weight. Plain clay tiles are covered by BS 402 and concrete tiles are covered by BS 473 and 550.

Plain tiling in common with other forms of tiling provides an effective barrier to rain and snow penetration but wind is able to penetrate into the building through the gaps between the tiling units, therefore a barrier in the form of boarding or sheeting is placed over the roof carcass before the battens on which the tiles are to be hung are fixed.

The rule for plain tiling is that there must always be at least two thicknesses of tiles covering any part of the roof and bonded so that no 'vertical' joint is immediately over a 'vertical' joint in the course below. To enable this rule to be maintained shorter length tiles are required at the eaves and the ridge, each alternate course is commenced with a wider tile of one-and-a-half tile widths. The apex or ridge is capped with a special tile bedded in cement mortar over the general tile surface. The hips can be covered with a ridge tile in which case the plain tiling is laid underneath and mitred on top of the hip or alternatively a special bonnet tile can be used where the plain tiles bond with the edges of the bonnet tiles. Valleys can be formed by using special tiles, mitred plain tiles or forming an open gutter with a durable material such as lead. The verge at the gable end can be formed by bedding plain tiles face down on the gable wall as an undercloak and bedding the plain tiles in cement mortar on the upper surface of the undercloak. The verge tiling should overhang its support by at least 50 mm. Abutments are made watertight by dressing a flashing over the upper surface of tiling between which is sandwiched a soaker. The soaker in effect forms a gutter. An alternative method is to form a cement fillet on top of the tiled surface but this method sometimes fails by the cement shrinking away from the surface of the wall.

The support or fixing battens are of softwood extended over and fixed to at least three rafters, the spacing or gauge being determined by the lap given to the tiles thus:

$$\text{Gauge} = \frac{\text{length of tile} - \text{lap}}{2}$$

$$\text{Gauge} = \frac{265 - 65}{2} = 100 \text{ mm}.$$

Plain tiles are fixed with two galvanised nails to each tile in every fourth or fifth course.

Details of plain tiles, fittings and methods of laying are shown in Figs. II.48 to II.52.

SINGLE LAP TILING

Single lap tiles are laid with overlapping side joints to a minimum pitch of 35° and are not bonded like the butt jointed single lap plain tiles; this gives an overall reduction in weight since less tiles are used. A common form of single lap tile is the pantile which has opposite corners mitred to overcome the problem of four tile thicknesses at the corners (see Fig. II.53). The pantile is a larger unit than the plain tile and is best employed on large roofs with gabled ends since the formation of hips and valleys is difficult and expensive. Other forms of single lap tiling are Roman tiling, Spanish tiling and interlocking tiling. The latter types are produced in both concrete and clay and have one or two grooves in the overlapping edge to give greater resistance to wind penetration and can generally be laid to lower pitches than other forms of tiling (see Fig. II.53).

Slating

Slate is a naturally dense material which can be split into thin sheets and used to provide a suitable covering to a pitched roof. Slates are laid to the same basic principles as double lap

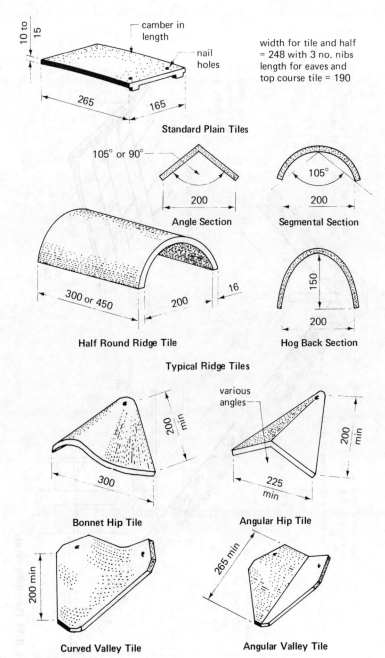

10 to 15

camber in length

nail holes

265   165

width for tile and half = 248 with 3 no. nibs
length for eaves and top course tile = 190

Standard Plain Tiles

105° or 90°

200

Angle Section

105°

200

Segmental Section

300 or 450   200   16

Half Round Ridge Tile

150

200

Hog Back Section

Typical Ridge Tiles

200 min

300

Bonnet Hip Tile

various angles

200 min

225 min

Angular Hip Tile

200 min

Curved Valley Tile

265 min

Angular Valley Tile

Fig. II.48 Standard plain tiles and fittings

115

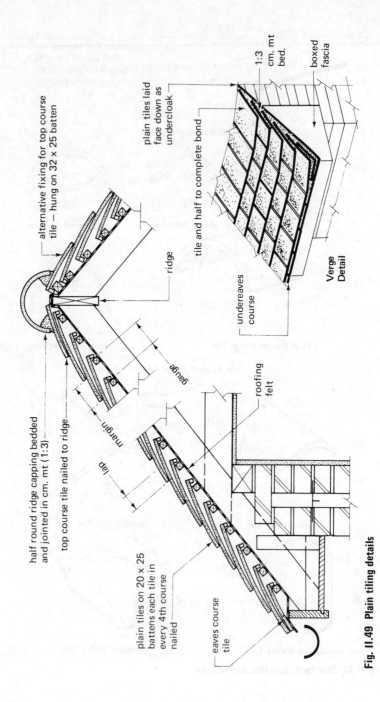

alternative fixing for top course tile – hung on 32 × 25 batten

1:3 cm. mt bed.

boxed fascia

plain tiles laid face down as undercloak

tile and half to complete bond

ridge

half round ridge capping bedded and jointed in cm. mt (1:3)

top course tile nailed to ridge

gauge

margin

lap

Verge Detail

undereaves course

roofing felt

plain tiles on 20 × 25 battens each tile in every 4th course nailed

eaves course tile

Fig. II.49 Plain tiling details

116

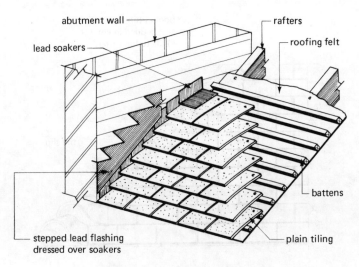

Typical Abutment Detail (see also Fig. II-55)

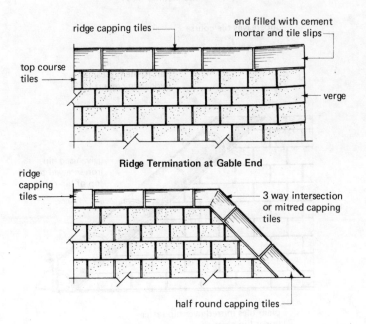

Ridge Termination at Gable End

Ridge Junction with Hipped End

Fig. II.50 Abutment and ridge details

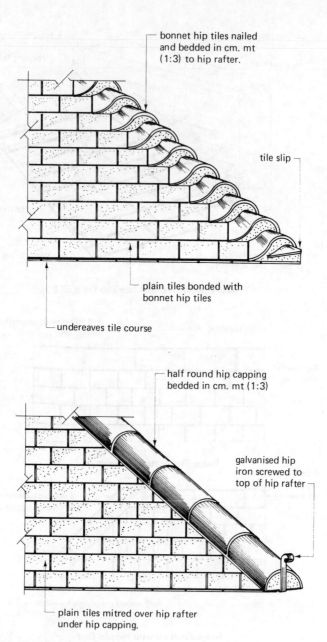

Fig. II.51 Hip treatments

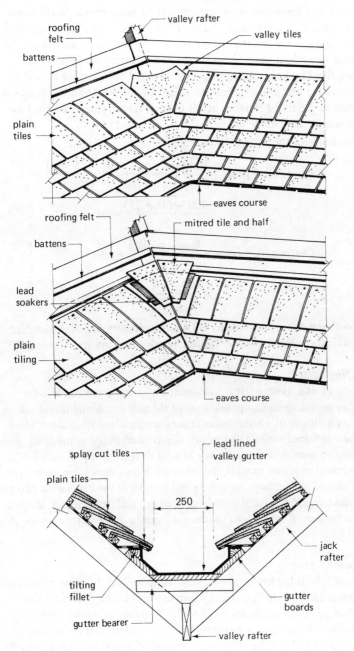

Fig. II.52 Valley treatments

Labels in figure:
- valley rafter
- valley tiles
- roofing felt
- battens
- plain tiles
- eaves course
- roofing felt
- mitred tile and half
- battens
- lead soakers
- plain tiling
- eaves course
- splay cut tiles
- lead lined valley gutter
- plain tiles
- 250
- jack rafter
- tilting fillet
- gutter boards
- gutter bearer
- valley rafter

roofing tiles except that every slate should be twice nailed. Slates come mainly from Wales, Cornwall and the Lake District and are cut to a wide variety of sizes—the Westmorland slates are harder to cut and are usually supplied in random sizes. Slates can be laid to a minimum pitch of 25° and are fixed by head nailing or centre nailing. Centre nailing is used to overcome the problem of vibration caused by the wind and tending to snap the slate at the fixing if nailed at the head, it is used mainly on the long slates and pitches below 35°.

The gauge of the battens is calculated thus:

$$\text{Head nailed gauge} = \frac{\text{length of slate} - (\text{lap} + 25 \text{ mm})}{2}$$

$$= \frac{400 - (75 + 25)}{2} = 150 \text{ mm}$$

$$\text{Centre nailed gauge} = \frac{\text{length of slate} - \text{lap}}{2}$$

$$= \frac{400 - 76}{2} = 162 \text{ mm}.$$

Roofing slates are covered by BS 680 which gives details of standard sizes, thicknesses and quality. Typical details of slating are shown in Fig. II.54.

Roofing felts

The object of a roofing felt is to keep out dust and the wind, also to assist in providing an acceptable level of thermal and sound insulation. It consists basically of a bituminous impregnated matted fibre sheet which can be reinforced with a layer of jute hessian embedded in the coating on one side to overcome the tendency of felts to tear readily. Roofing felts are supplied in rolls 1 m wide and 10 or 20 m long depending upon type. They should be laid over the rafters and parallel to the eaves with 150 mm laps and temporarily fixed with large head felt nails until finally secured by the battens. Roofing felts should conform to the requirements of BS 747.

Counter battens

If a roof is boarded before the roofing felt is applied the tiling battens will provide ledges on which dirt and damp can collect. To overcome this problem counter battens are fixed to the boarding over the rafter positions to form a cavity between the tiling battens and the boarding. Boarding a roof over the rafters is a method of providing a wind barrier and adding to the thermal insulation properties but is seldom used in new work because of the high cost.

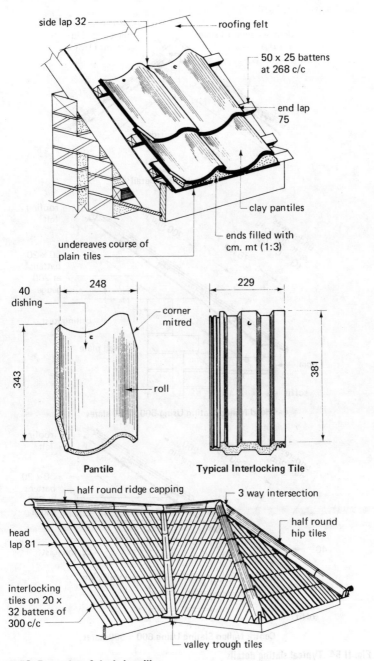

Fig. II.53 Examples of single lap tiling

121

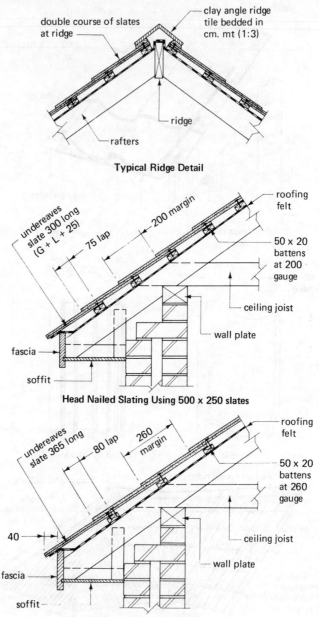

Typical Ridge Detail

Head Nailed Slating Using 500 x 250 slates

Centre Nailed Slating Using 600 x 300 slates

Fig. II.54 Typical slating details

13
Fireplaces, chimneys and flues

The traditional method of providing heating in a domestic building is the open fire burning coal and/or wood but because of its low efficiency compared to modern heating appliances its use as the primary source of heating is declining. It must be noted that all combustible fuels, such as coal, smokeless fuels, oils, gas and wood, require some means of conveying the products of combustion away from the appliance or fireplace to the open air.

Terminology

Fireplace: this is the area in which combustion of the fuel takes place. It may be in the form of an open space with a fret or grate in which wood or coal is burnt, or a free standing appliance such as a slow combustion stove, room heater, oil burning appliance, gas fire or gas boiler.

Chimney: the fabric surrounding the flue and providing it with the necessary strength and protection.

Flue: this is strictly a void through which the products of combustion pass. It is formed by lining the inside of a chimney with a suitable lining material to give protection to the chimney fabric from the products of combustion, as well as forming a flue of the correct shape and size to suit the type of fuel and appliance being used.

Building Regulations J1 to J3 are concerned with heat-producing appliances and requires that they are installed with an adequate air supply for efficient working, provided with a suitable means of discharging the products of combustion to the outside air and installed so as to reduce the risk of the building catching fire. Approved Document J gives practical guidance to satisfy these regulations and makes particular reference to solid-fuel and oil-burning appliances with a rated output of up to 45 kW and it is this class with which a basic technology course is concerned.

When a fuel is burnt to provide heat it must have an adequate air supply to provide the necessary oxygen for combustion to take place. The air supply is drawn through the firebed in the case of solid fuels and injected in the case of oil burning appliances. A secondary supply of air is also drawn over the flames in both cases. In rooms where an open fire is used the removal of air induces ventilation and this helps to combat condensation. The air used in combustion is drawn through the appliance by suction provided by the flue and this is affected by the temperature, pressure, volume, cross-sectional area and surface lining of the flue.

To improve the efficiency of a flue it is advantageous for the chimney to be situated on an internal wall since this will reduce the heat losses to the outside air, also the flue gas temperature will be maintained and this will, in turn, reduce the risk of condensation within the flue. The condensation could cause corrosion of the chimney fabric due to the sulphur compounds which are a by product of the processes of combustion.

Chimneys

A chimney should be built as vertical as is practicable to give maximum flue gas flow—if bends are required the angle should lie between 45 and 30°. If an appliance is connected to a chimney the flue pipe should be as short as possible and continuous with an access for cleaning included at the base of the main flue. The chimney should be terminated above the roof so that it complies with the Building Regulations and is unaffected by adverse pressures which occur generally on the windward side of a roof and may cause down draughts. The use of a chimney terminal or pot could be to transfer the rectangular flue section to a circular cross-section giving better flue flow properties or to provide a cover to protect the flue from the entry of rain or birds. A terminal should be fixed so that it does not impede in any way the flow of flue gasses.

BUILDING REGULATIONS 1985

Heat-producing appliances are covered by Part J supported by Approved Document J which has two sections each divided into three parts. The main recommendations for appliances with a rated output of up to 45 kW together with typical construction details are shown in Figs II.55–II.58.

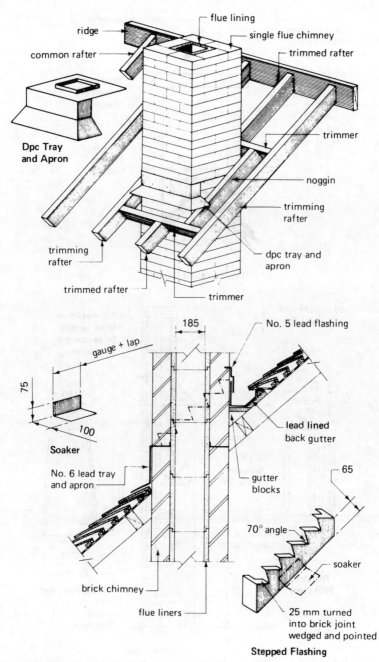

ridge

common rafter

flue lining

single flue chimney

trimmed rafter

trimmer

noggin

trimming rafter

dpc tray and apron

Dpc Tray and Apron

trimming rafter

trimmed rafter

trimmer

185

gauge + lap

75

100

Soaker

No. 6 lead tray and apron

brick chimney

flue liners

No. 5 lead flashing

lead lined back gutter

gutter blocks

65

70° angle

soaker

25 mm turned into brick joint wedged and pointed

Stepped Flashing

Fig. II.55 Chimney construction

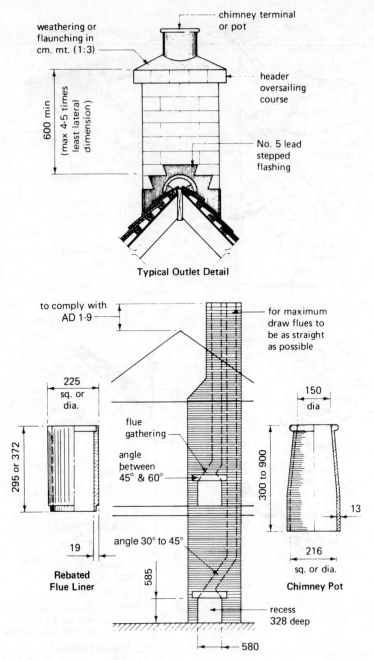

Typical Outlet Detail

chimney terminal or pot

weathering or flaunching in cm. mt. (1:3)

header oversailing course

600 min (max 4·5 times least lateral dimension)

No. 5 lead stepped flashing

to comply with AD 1·9

for maximum draw flues to be as straight as possible

225 sq. or dia.

flue gathering

angle between 45° & 60°

295 or 372

150 dia

300 to 900

13

19

angle 30° to 45°

216 sq. or dia.

585

Chimney Pot

Rebated Flue Liner

recess 328 deep

580

Fig. II.56 Typical flue construction

126

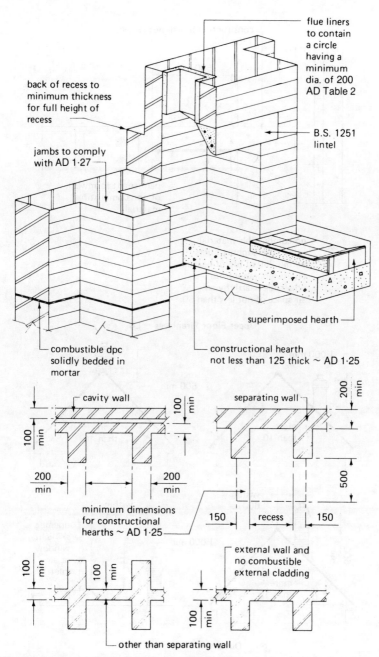

Fig. II.57 Fireplaces and Approved Document J

127

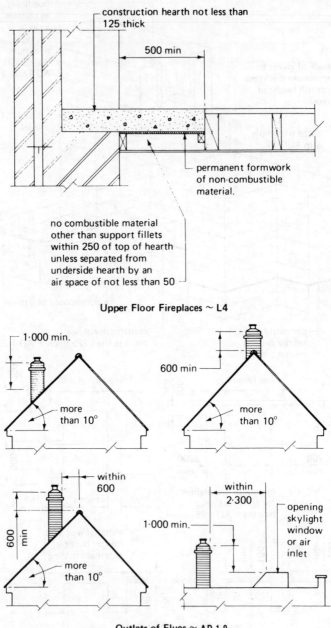

construction hearth not less than 125 thick

500 min

permanent formwork of non-combustible material.

no combustible material other than support fillets within 250 of top of hearth unless separated from underside hearth by an air space of not less than 50

Upper Floor Fireplaces ~ L4

1·000 min.

more than 10°

600 min

more than 10°

within 600

600 min

more than 10°

within 2·300

opening skylight window or air inlet

1·000 min.

Outlets of Flues ~ AD 1.9

Fig. II.58 Fireplaces and outlets– Approved Document J

Firebacks

The functions of a fireback are to contain the burning fuel, to prevent the heat of the fire from damaging the wall behind it and to radiate heat from the fire into the room. The space behind a fireback should be of a solid material though not a strong mix and not composed of loose rubble, the recommended mix is 1 : 2 : 4 lime, sand and broken brick, alternatively an insulating mix of 5 parts of vermiculite to 1 part of cement could be used. The temperature encountered by a fireback will be high, therefore the problem of expansion and subsequent contraction must be considered. The fireback should preferably be in two or more parts since the lower half will become hotter than the upper half. Multi-piece firebacks also have the advantage of being easier to fit. It is also good practice to line the rear of a fireback with corrugated paper or a similar material which will eventually smoulder away leaving a small expansion gap at the back of the fireback.

Throat

The size and shape of the throat above the fireplace opening is of the utmost importance. A fireplace with an unrestricted outlet to the fire would create unpleasant draughts by drawing an unnecessary amount of air through the flue and reduce the efficiency of the fire by allowing too much heat to escape into the flue. A throat restriction of 100 mm will give reasonable efficiency without making chimney sweeping impossible (see Fig. II.59).

Surrounds

This is the façade of a fireplace and its main function is an attractive appearance. It can be of precast concrete with an applied finish such as tiles or built *in situ* from small brickettes or natural stones. If it is precast it is usually supplied in two pieces, the front and the hearth. The front is fixed by screws through lugs cast into the edges of the surround and placed against a 25 mm wide non-combustible cord or rope around the fireplace opening to allow for expansion and contraction. The hearth should be bedded evenly on the constructional hearth with at least 10 mm of 1 : 1 : 8 cement : lime : sand mix (see Fig. II.59).

DEEP ASHPIT FIRE

This is a low front fire with a fret or grate being at or just below hearth level. A pit is constructed below the grate to house a large ashpan capable of holding several days ash. The air for combustion is introduced through the ashpit and grate by means of a 75 mm diameter duct at the end of which is an air control regulator. If the floor is of suspended construction the air duct would be terminated after passing through the fender wall, whereas with a solid floor the duct must pass under the floor and be terminated beyond the external wall, the end being suitably guarded against the entry of vermin.

BACK BOILER

This an open fire of conventional design with part of the fireback replaced by a small boiler with a boiler flue and a control damper. It is used to supply hot water primarily for domestic use.

OPEN FIRE CONVECTORS

These are designed to increase the efficiency of an open fire by passing back into the room warm air as well as the radiated heat from the burning fuel. The open convector is a self-contained unit consisting of a cast-iron box containing the grate and fireback forming a convection chamber. The air in the chamber heats up and flows into the room by convection currents moving the air up and out of the opening at the top of the unit.

ROOM HEATERS

These are similar appliances to open fire convectors but they are designed to burn smokeless fuels and operate as a closed unit. In some models the strip glass front is, in fact, a door and can therefore be opened to operate as an open fire. Room heaters are either freestanding, that is, fixed in front of the surround with a plate reducing the size of the fireplace opening to the required flue pipe size, or inset into the fireplace recess where the chimney flue aperture is reduced by a plate to receive the flue pipe of the appliance.

INDEPENDENT BOILERS

The function of this appliance is to heat water whether it is for a central heating system or merely to provide hot water for domestic use. It is generally fixed in the kitchen because of the convenience to the plumbing pipe runs and to utilise the background heat emitted. It must discharge into its own flue since it is a slow combustion appliance and the fumes emitted could, if joined to a common flue, cause a health hazard in other rooms of the building. Adequate access must be made for sweeping the flue and removing the fly ash that will accumulate with a smokeless fuel (see Fig. II.59).

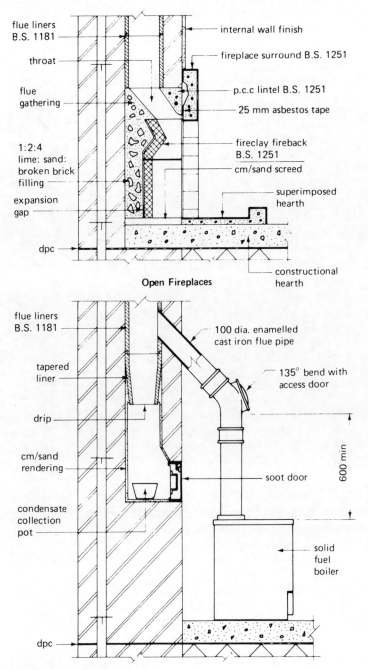

Fig. II.59 Typical domestic fireplaces

131

Part III
Finishes and fittings

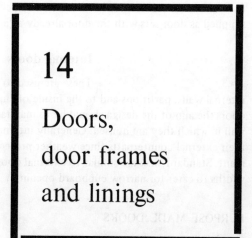

14
Doors,
door frames
and linings

Doors

A door is a screen used to seal an opening into a building or between rooms within a building. It can be made of timber, glass, metal or plastic or any combination of these materials. Doors can be designed to swing from one edge, slide, slide and fold or roll to close an opening. The doors to be considered in this book are those made of timber and those made of timber and glass which are hung so that they swing from one edge. All doors may be classified by their position in a building, by their function or by their method of construction.

External doors

These are used to close the access to the interior of a building and provide a measure of security. They need to be weather resistant since in general they are exposed to the elements, this resistance is provided by the thickness, stability and durability of the construction and materials used together with protective coatings of paint or polish. The external walls of a building are designed to give the interior of a building a degree of thermal and sound insulation, doors in such walls should therefore be constructed, as far as practicable, to maintain the insulation properties of the external enclosure.

The standard sizes for external timber doors are 1981 mm high x 762 or 838 mm wide x 45 mm thick which is a metric conversion of the old Imperial door sizes. Metric doors are produced so that, together with the frame, they fit into a modular coordinated opening size and are usually supplied as door sets with the door already attached or hung in the frame.

Internal doors

These are used to close the access through internal walls, partitions and to the inside of cupboards. As with external doors the aim of the design should be to maintain the properties of the wall in which they are housed. Generally internal doors are thinner than their external counterparts since weather protection is no longer a requirement. Standard sizes are similar to external doors but with a wider range of widths to cater for narrow cupboard openings.

PURPOSE MADE DOORS

The design and construction of these doors is usually based on BS 459 for standard doors but are made to non-standard sizes, shapes or designs. Most door manufacturers produce a range of non-standard doors which are often ornate and are used mainly for the front elevation doors of domestic buildings. Purpose made doors are also used in buildings such as banks, civic buildings, shops, theatres and hotels to blend with or emphasise the external façade design or internal decor (see Fig. III.1).

METHODS OF CONSTRUCTION

The British Standard Code of Practice 151 for wood doors and frames covers all aspects of door and frame construction. A British Standard for special door formats is also published:

BS 459 Part 3: Fire-check doors.
       Part 4: Matchboarded doors.
CP 151: all wood doors and frames.

Standard doors are used extensively since they are mass produced to known requirements, are readily available from stock and are cheaper than purpose made doors.

Panelled and glazed wood doors

The Code of Practice gives a wide variety of types all of which are based upon the one, two, three or four panel format. They are constructed of timber which should be in accordance with the recommendations of BS 1186 with plywood or glass panels. External doors with panels of plywood should be constructed using an external quality plywood (see Fig. III.2).

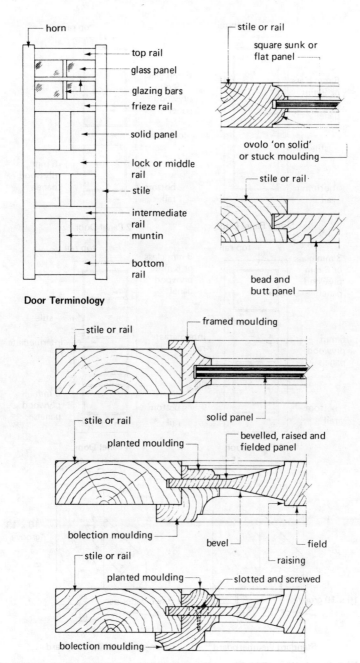

Door Terminology

horn

top rail

glass panel

glazing bars

frieze rail

solid panel

lock or middle rail

stile

intermediate rail

muntin

bottom rail

stile or rail

square sunk or flat panel

ovolo 'on solid' or stuck moulding

stile or rail

bead and butt panel

stile or rail

framed moulding

solid panel

stile or rail

planted moulding

bevelled, raised and fielded panel

bolection moulding

bevel

raising

field

stile or rail

planted moulding

slotted and screwed

bolection moulding

Fig. III.1 Purpose made doors and mouldings

137

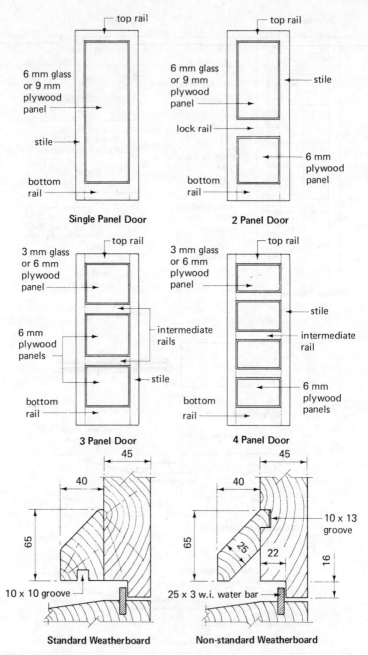

Fig. III.2 Standard panelled doors and weatherboards

The joints used in framing the doors can be a dowelled joint or a mortise and tenon joint. The dowelled joint is considered superior to the mortise and tenon joint and is cheaper when used in the mass production of standard doors. Bottom and lock rails have three dowels, top rails have two dowels and intermediate rails have a single dowel connection (see Fig. III.3). The plywood panels are framed into grooves with closely fitting sides with a movement allowance within the depth of the groove of 2 mm. The mouldings at the rail intersections are scribed, whereas the loose glazing beads are mitred. Weatherboards for use on external doors can be supplied to fit onto the bottom rail of the door which can also be rebated to close over a water bar (see Fig. III.2).

Flush doors

This type of door is very popular with both the designer and the occupier—it has a plain face which is easy to clean and decorate, it is also free of the mouldings which collect dust. Flush doors can be faced with hardboard, plywood or a plastic laminate and by using a thin sheet veneer of good quality timber the appearance of high class joinery can be created.

The Code of Practice specifies the requirements for flush doors but leaves the method of construction to the manufacturer which gives him complete freedom in his design; therefore the forms of flush door construction are many and varied but basically they can be considered as either skeleton core doors or solid core doors. The former consists of an outer frame with small section intermediate members over which is fixed the facing material. The facing has a tendency to deflect between the core members and this can be very noticeable on the surface especially if the facing is coated with gloss paint. Solid core doors are designed to overcome this problem and at the same time improve the sound insulation properties of flush doors. Solid doors of suitably faced block or lamin board are available for internal and external use. Another method of construction is to infill the voids created by a skeleton core with a lightweight material such as foamed plastic which will give support to the facings but will not add appreciably to the weight of the door.

The facings of flush doors are very vulnerable to damage at the edges, therefore a lipping of solid material should be fixed to at least the vertical edges (good class doors have lippings on all four edges).

Small glazed observation panels can be incorporated in flush doors when the glass panel is secured by loose fixing beads (see Figs. III.4 and III.5).

Fire-check flush doors

These doors provide an effective barrier to the passage of fire for the time designated by their type but to achieve this they must be used in

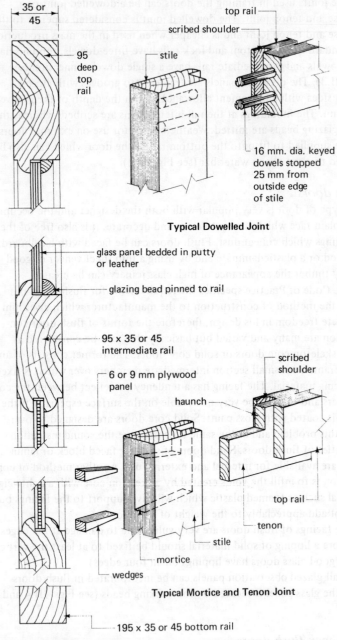

Fig. III.3 Panelled door details

Labels in figure:

- 35 or 45
- 95 deep top rail
- stile
- top rail
- scribed shoulder
- 16 mm. dia. keyed dowels stopped 25 mm from outside edge of stile
- **Typical Dowelled Joint**
- glass panel bedded in putty or leather
- glazing bead pinned to rail
- 95 x 35 or 45 intermediate rail
- 6 or 9 mm plywood panel
- scribed shoulder
- haunch
- top rail
- tenon
- stile
- mortice
- wedges
- **Typical Mortice and Tenon Joint**
- 195 x 35 or 45 bottom rail

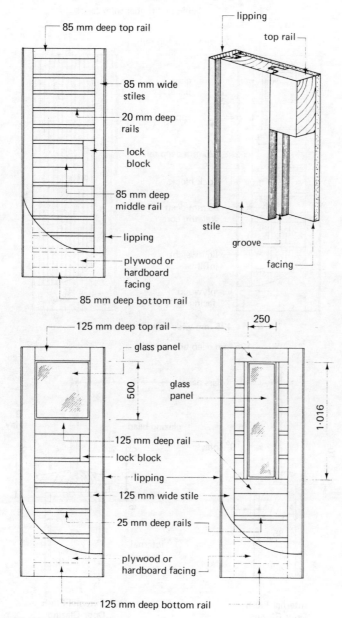

85 mm deep top rail

85 mm wide stiles

20 mm deep rails

lock block

85 mm deep middle rail

lipping

plywood or hardboard facing

85 mm deep bottom rail

lipping

top rail

stile

groove

facing

125 mm deep top rail

glass panel

500

glass panel

250

1·016

125 mm deep rail

lock block

lipping

125 mm wide stile

25 mm deep rails

plywood or hardboard facing

125 mm deep bottom rail

Fig. III.4 Skeleton core flush doors

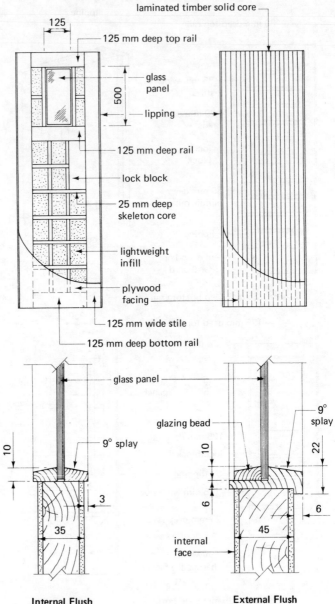

laminated timber solid core

125

125 mm deep top rail

glass panel

500

lipping

125 mm deep rail

lock block

25 mm deep skeleton core

lightweight infill

plywood facing

125 mm wide stile

125 mm deep bottom rail

glass panel

10

9° splay

3

35

Internal Flush Door Glazing

glass panel

glazing bead

9° splay

10

22

6

6

45

internal face

External Flush Door Glazing

Fig. III.5 Solid core doors

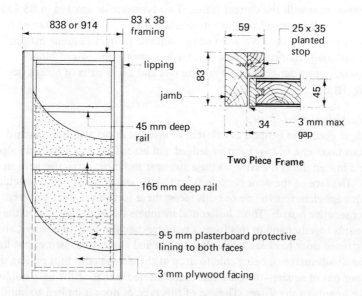

83 x 38 framing

838 or 914

lipping

jamb

83

45 mm deep rail

165 mm deep rail

9·5 mm plasterboard protective lining to both faces

3 mm plywood facing

Two Piece Frame

59

25 x 35 planted stop

45

34

3 mm max gap

Half-hour Type Fire-check Door and Frame

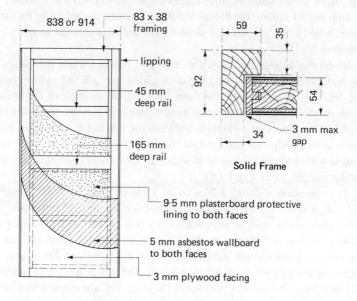

83 x 38 framing

838 or 914

lipping

45 deep rail

165 mm deep rail

9·5 mm plasterboard protective lining to both faces

5 mm asbestos wallboard to both faces

3 mm plywood facing

92

35

59

54

34

3 mm max gap

Solid Frame

One Hour Type Fire-check Door and Frame

Fig. III.6 Fire-check doors and frames

143

conjunction with the correct frame. Two types are designated in BS 459. namely half-hour and one-hour resistance. This resistance is obtained by placing beneath the plywood facing a suitable protective lining material or materials. Half-hour type doors are hung using one pair of hinges whereas one-hour type doors require one and a half pairs of hinges (see Fig. III.6).

Matchboarded doors

These doors can be used as external or internal doors and as a standard door takes one of two forms, a ledged and braced door or a framed ledged and braced door, the latter being a stronger and more attractive version.

The face of the door is made from tongue and grooved boarding which has edge chamfers to one or both faces; these form a vee joint between consecutive boards. Three horizontal members called 'ledges' clamp the boards together and in this form a non-standard door called a ledged and battened door has been made. It is simple and cheap to construct but has the disadvantage of being able to drop at the closing edge thus pulling the door out of square—the only resistance offered is that of the nails holding the boards to the ledges. The use of this type of door is limited to build-ings such as sheds, outhouses and to small units like trapdoors (see Fig. III.7). In the standard door braces are added to resist the tendency to drop out of square; the braces are fixed between the ledges so that they are parallel to one another and slope downwards towards the hanging edge (see Fig. III.7).

In the second standard type a mortise and tenoned frame surrounds the matchboarded panel giving the door added strength and rigidity (see Fig. III.7). If wide doors of this form are required the angle of the braces becomes too low to be of value as an effective restraint and the brace must therefore be framed as a diagonal between the top and bottom rails. Wide doors of this design are not covered by the British Standard but are often used in pairs as garage doors or as wide entrance doors to workshops and similar buildings (see Fig. III.8).

The operation of fixing a door to its frame or lining is termed hanging and entails removing the protective horns from the top and bottom of the stiles; planing the stiles to reduce the door to a suitable width; cutting and planing the top and bottom to the desired height; marking out and fitting the butts or hinges which attach the door to the frame and fitting any locks and door furniture which is required. The hinges should be positioned 225 mm from the top and bottom of the door and where one and a half pairs of hinges are specified for heavy doors the third hinge is positioned midway between the bottom and top hinges.

A door irrespective of the soundness of its construction will deteriorate if improperly treated during transportation, storage and after hanging. It

144

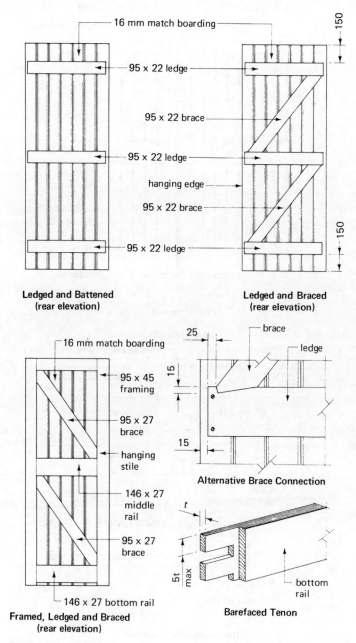

Ledged and Battened
(rear elevation)

Ledged and Braced
(rear elevation)

16 mm match boarding

95 x 22 ledge

95 x 22 brace

95 x 22 ledge

hanging edge

95 x 22 brace

95 x 22 ledge

150

150

16 mm match boarding

95 x 45 framing

95 x 27 brace

hanging stile

146 x 27 middle rail

95 x 27 brace

146 x 27 bottom rail

Framed, Ledged and Braced
(rear elevation)

25

15

15

brace

ledge

Alternative Brace Connection

t

5t max

bottom rail

Barefaced Tenon

Fig. III.7 Matchboarded doors

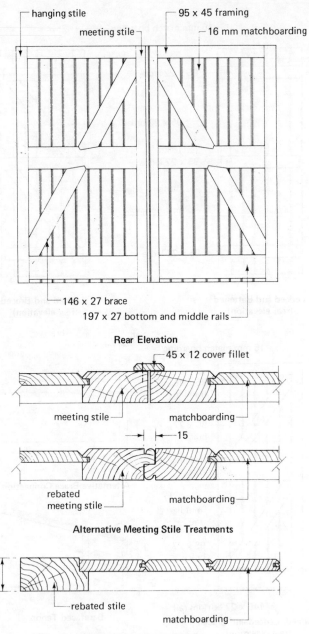

hanging stile

meeting stile

95 x 45 framing

16 mm matchboarding

146 x 27 brace

197 x 27 bottom and middle rails

Rear Elevation

45 x 12 cover fillet

meeting stile

matchboarding

15

rebated
meeting stile

matchboarding

Alternative Meeting Stile Treatments

45

rebated stile

matchboarding

Alternative Stile Treatment

Fig. III.8 Matchboarded double doors

should receive a wood priming coat of paint before or immediately after delivery, be stored in the dry and in a flat position so that it does not twist—it should also receive the finishing coats of paint as soon as practicable after hanging.

Frames and linings

A door frame or lining is attached to the opening in which a door is to be fitted, it provides a surround for the door and is the member to which a door is fixed or hung. Door sets consisting of a storey heigh. frame with a solid or glazed panel over the door head are also available; these come supplied with the door ready hung on lift off hinges (see Fig. III.10).

TIMBER DOOR FRAMES

These are made from rectangular section timber in which a rebate is formed or to which a planted door stop is fixed to provide the housing for the door. Generally a door frame is approximately twice as wide as its thickness plus the stop. Frames are used for most external doors, for heavy doors and for doors situated in thin non-load bearing partitions.

A timber door frame consists of three or four members—namely the head, two posts or jambs and a sill or threshold. The members can be joined together by wedged mortise and tenon joints, combed joints or mortise and tenon joints pinned with a metal star shaped dowel or a round timber dowel. All joints should have either a coating of adhesive or a coating of a lead based paint (see Fig. III.9).

Door frames which do not have a sill are fitted with mild steel dowels driven into the base of the jambs and cast into the floor slab or alternatively grouted into pre-formed pockets as a means of securing the feet of the frame to the floor. If the frame is in an exposed position it is advisable to sit the feet of the jambs on a damp-proof pad such as lead or bituminous felt, to prevent moisture soaking into the frame and creating the conditions for fungi attack.

Door frames fitted with a sill are designed for one of two conditions:

1. Doors opening out.
2. Doors opening in.

In both cases the sill must be designed to prevent the entry of rain and wind under the bottom edge of the door. Doors opening out close onto a rebate in the sill whereas doors opening in have a rebated bottom rail and close over a water bar set into the sill (see Fig. III.9).

147

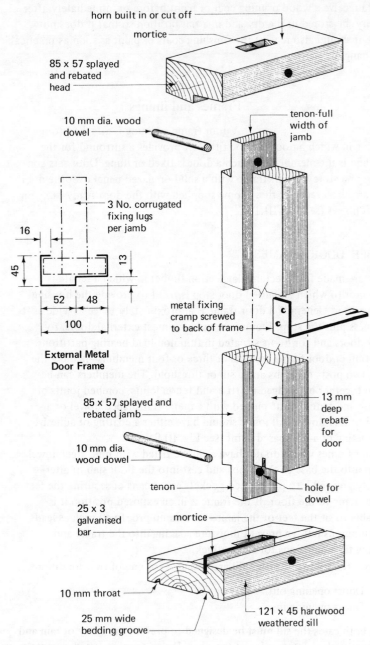

horn built in or cut off

mortice

85 x 57 splayed
and rebated
head

tenon-full
width of
jamb

10 mm dia. wood
dowel

3 No. corrugated
fixing lugs
per jamb

16

45

13

52   48

100

**External Metal
Door Frame**

metal fixing
cramp screwed
to back of frame

85 x 57 splayed and
rebated jamb

13 mm
deep
rebate
for
door

10 mm dia.
wood dowel

tenon

hole for
dowel

25 x 3
galvanised
bar

mortice

10 mm throat

25 mm wide
bedding groove

121 x 45 hardwood
weathered sill

Fig. III.9 Door frames

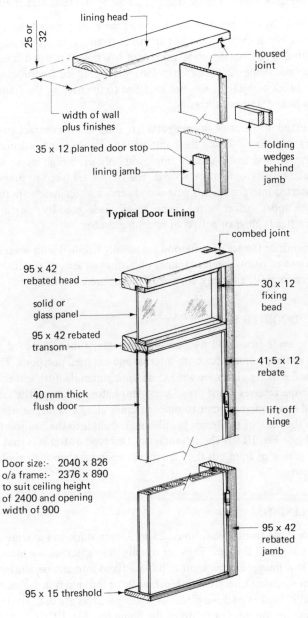

Typical Door Lining

lining head

25 or 32

width of wall plus finishes

35 x 12 planted door stop

lining jamb

housed joint

folding wedges behind jamb

combed joint

95 x 42 rebated head

solid or glass panel

95 x 42 rebated transom

40 mm thick flush door

30 x 12 fixing bead

41·5 x 12 rebate

lift off hinge

Door size:- 2040 x 826
o/a frame:- 2376 x 890
to suit ceiling height
of 2400 and opening
width of 900

95 x 42 rebated jamb

95 x 15 threshold

Typical Door Set

Fig. III.10 Door linings and door sets

Timber door frames can be fixed to a wall by the following methods:

(*a*) Built into the brick or block wall as the work proceeds by using 'L' shaped ties or cramps. The ties are made from galvanised wrought steel with one end turned up 50 mm, with two holes for wood screws, the other end being 125 or 225 mm long and fish-tailed for building into brick or block bed joints. The ties are fixed to the back of the frame for building in at 450 mm centres.

(*b*) Fixed into a brick opening at a late stage in the contract to prevent damage to the frame during the construction period. This is a more expensive method and is usually employed only when high class joinery using good quality timber is involved. The frames are fixed to timber plugs inserted into the reveals with wood screws whose heads are sunk below the upper surface of the frame; this is made good by inserting over the screw heads plugs or pellets of matching timber.

Timber door frames of softwood are usually finished with several applications of paint, whereas frames of hardwood are either polished or oiled. Frames with a factory coating of plastic are also available.

METAL DOOR FRAMES

These are made from mild steel pressed into one of three standard profiles and are suitable for both internal and external positions. The hinges and striking plates are welded on during manufacture and the whole frame receives a rust-proof treatment before delivery. The frames are fixed in a similar manner to timber frames using a tie or lug which fits into the back of the frame profile and is built into the bed joints of the wall (see Fig. III.9). The advantage of this type of frame is that they will not shrink or warp but they are more expensive than their timber counterparts.

DOOR LININGS

These are made from timber board 25 or 32 mm thick and as wide as the wall plus any wall finishes. They are usually only specified for internal doors. Door linings are not built in but are fixed into an opening by nailing or screwing directly into block walls or into plugs in the case of brick walls. Timber packing pieces or folding wedges are used to straighten and plumb up the sides or jambs of the lining (see Fig. III.10).

15
Windows,
glass and glazing

The primary function of a window is to
provide a means for admission of natural daylight to the interior of a
building. A window can also serve as a means of providing the necessary
ventilation of dwellings, as required under Building Regulation F1, by
including into the window design opening lights.

Windows, like doors, can be made from a variety of materials or a
combination of these materials such as timber, metal and plastic. They
can also be designed to operate in various ways by arranging for the sashes
to slide, pivot or swing, by being hung to one of the frame members. This
latter arrangement is known as a casement window and it is this form
which is studied in this volume.

BUILDING REGULATIONS

Approved Document F deals with the ventilation of habitable rooms and
for the purposes of Regulation F1 a room used for kitchen or bathroom
is to be ventilated as a habitable room. A habitable room must have
ventilation openings unless it is adequately ventilated by mechanical
means. There is no definition in the Building Regulations of adequate
mechanical means but it is generally recommended that one to three
air changes per hour would be a reasonable ventilation standard.

A ventilation opening is any part of a window or any hinged panel,
adjustable louvre or other means of ventilation which opens directly to
the external air, but excludes any opening associated with a mechanically
operated system. A door, if it opens directly to the external air, can be

classed as a ventilation opening if it has an opening ventilator with an area of not less than 10 000 mm^2 or if it is situated in a room which contains one or more ventilation openings whose total area is not less than 10 000 mm^2.

The basic requirements for ventilation openings are:

1. At least one ventilation opening must exceed one-twentieth of the floor area of the room it serves.
2. Some part of the ventilation opening must be not less than 1 750 mm above the floor level.
3. Rooms with an enclosed veranda or conservatory must have ventilation opening(s) whose total area is not less than one-twentieth of the combined floor areas.
4. Any larder can be ventilated to the external air and if this is achieved by using a window it must have ventilation opening whose total area is not less than 85 000 mm^2 and must be fitted with a durable fly-proof screen. Refrigeration is an acceptable alternative to natural ventilation.

Traditional casement windows

Figure III.11 shows a typical arrangement and details of this type of window. A wide range of designs can be produced by using various combinations of the members, the only limiting factor being the size of glass pane relevant to its thickness.

The general arrangement of the framing is important, heads and sills always extend across the full width of the frame and in many cases have projecting horns for building into the wall. The jambs and mullions span between the head and sill; these are joined to them by a wedged or pinned mortise and tenon joint. This arrangement gives maximum strength since the vertical members will act as struts—it will also give a simple assembly process.

The traditional casement window frame has deep rebates to accommodate the full thickness of the sash, which is the term used for the framing of the opening ventilator. If fixed glazing or lights are required it is necessary to have a sash frame surround to the glass since the depth of rebate in the window frame is too great for direct glazing to the frame.

STANDARD WOOD CASEMENT WINDOWS

BS 644, Part 1, gives details of the quality, construction and design of a wide range of wood casement windows. The frames, sashes and ventlights are made from standard sections of softwood timbers arranged to give a variety in design and size. The sashes and ventlights are designed so that their edges

152

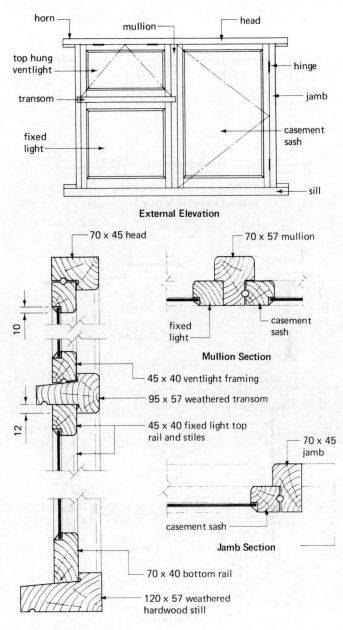

External Elevation

horn

mullion

head

top hung ventlight

hinge

transom

jamb

fixed light

casement sash

sill

70 x 45 head

70 x 57 mullion

fixed light

casement sash

Mullion Section

10

12

45 x 40 ventlight framing

95 x 57 weathered transom

45 x 40 fixed light top rail and stiles

70 x 45 jamb

casement sash

Jamb Section

70 x 40 bottom rail

120 x 57 weathered hardwood still

Vertical Section

Fig. III.11 Traditional timber casement window

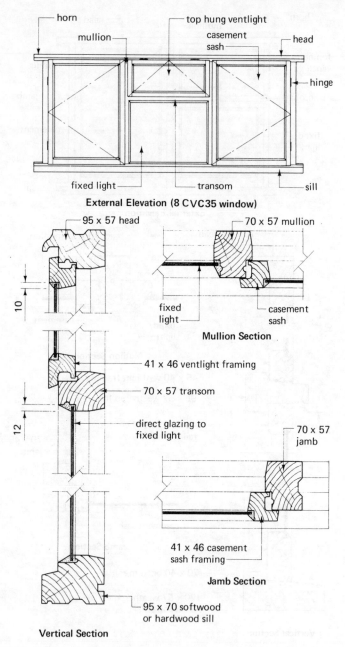

External Elevation (8 CVC35 window)

horn

mullion

top hung ventlight

casement sash

head

hinge

fixed light

transom

sill

95 x 57 head

70 x 57 mullion

10

fixed light

casement sash

Mullion Section

41 x 46 ventlight framing

70 x 57 transom

12

direct glazing to fixed light

70 x 57 jamb

41 x 46 casement sash framing

Jamb Section

95 x 70 softwood or hardwood sill

Vertical Section

Fig. III.12 Typical modified BS casement window

rebate over the external face of the frame to form a double barrier to the entry of wind and rain. The general construction is similar to that described for traditional casement windows and the fixing of the frame into walls follows that described for door frames.

Most joinery manufacturers produced a range of modified standard casement windows following the basic principles set out in BS 644 but with improved head, sill and sash sections. The range produced is based on a module for basic spaces of 300 mm giving the following lengths (in mm):

600; 900; 1 200; 1 800; 2 400.

Frame heights follow the same pattern with the exception of one half module (in mm):

600; 900; 1 050; 1 200; 1 500.

Window types are identified by a notation of figures and letters, for example, 4CV30 where:

4 = four width modules = 4 x 300 mm = 1 200 mm.
C = casement.
V = ventlight.
30 = three height modules = 3 x 300 mm = 900 mm.

For typical details see Fig. III.12.

STEEL CASEMENT WINDOWS

These windows are produced to conform with the recommendations of BS 6510 which gives details of construction, sections, sizes, composites and hardware. The standard range covers fixed lights, hung casements, pivot casements and doors. The lengths, in the main, conform to the basic space first preference of 300 mm giving the following range (in mm):

500; 600; 800; 900; 1 200; 1 500; 1 800.

Frame heights are based upon basic spaces for the preferred head and sill heights for public sector housing giving the following sizes (in mm):

200; 500; 700; 900; 1 100; 1 300; 1 500.

Steel windows, like wood windows, are identified by a notation of numbers and letters:

Prefix number: x 100 = basic space length.

Code letters: F = fixed light.
C = side hung casement opening out.

155

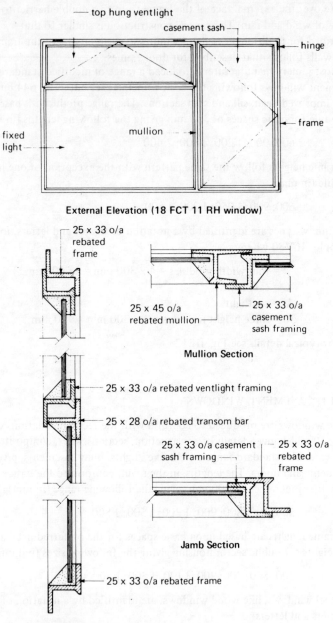

External Elevation (18 FCT 11 RH window)

top hung ventlight

casement sash

hinge

frame

fixed light

mullion

25 x 33 o/a rebated frame

25 x 45 o/a rebated mullion

25 x 33 o/a casement sash framing

Mullion Section

25 x 33 o/a rebated ventlight framing

25 x 28 o/a rebated transom bar

25 x 33 o/a casement sash framing

25 x 33 o/a rebated frame

Jamb Section

25 x 33 o/a rebated frame

Vertical Section

Fig. III.13 Typical steel window details

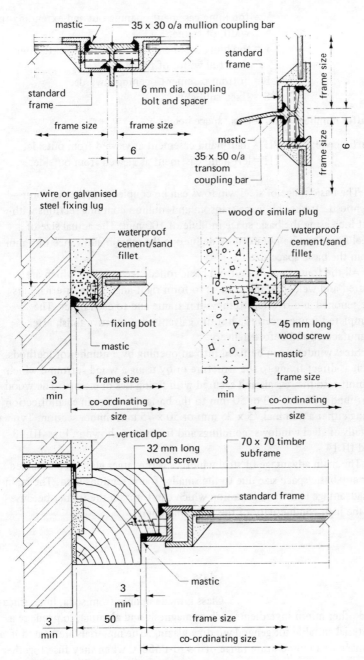

Fig. III.14 Steel window couplings and fixings

157

V = top hung casement opening out and extending full
width of frame.

T = top hung casement opening out and extending less
than full width of frame.

B = bottom casement opening inwards.

S = fixed sublight.

Suffix number: x 100 = basic space height.

Suffix code: RH = right-hand casement as viewed from outside.

LH = left-hand casement as viewed from outside.

The basic range of steel windows can be coupled together to form
composite frames by using transom and mullion coupling sections without increasing the basic space module of 100 mm. The actual size of a
steel frame can be obtained by deducting the margin allowance of 6 mm
from the basic space size.

All the frames are made from basic rolled steel sections which are
mitred and welded at the corners to form right-angles; the internal bars
are tenoned and rivetted to the outer frame and to each other. The
completed frame receives a hot dip galvanised protective finish after
manufacture and before delivery.

Steel windows can be fixed into an opening by a number of methods
such as direct fixing to the structure or by using a wood surround which
is built into the reveals and secured with fixing ties or cramps. The wood
surround will add 100 or 50 mm to the basic space size in each direction
using either a nominal 75 x 75 mm or 50 x 75 mm timber section. Typical
details of steel windows, couplings and fixings are shown in Figs. III.13
and III.14.

The main advantage of steel windows is the larger glass area obtained
for any basic space size due to the smaller frame sections used. The main
disadvantage is the condensation which can form on the frames because
of the high conductivity of the metal members.

Glass

Glass is made mainly from soda, lime, silica
and other minor ingredients such as magnesia and alumina, to produce a
material suitable for general window glazing. The materials are heated in a
furnace to a temperature range of 1 490-1 550°C when they fuse together
in a molten state—they are then formed into sheets by a process of drawing, floating or rolling.

DRAWN CLEAR SHEET GLASS

There are two principal methods of producing drawn clear sheet glass: the first is by vertical drawing from a pool of molten glass which when 1 m or so above the pool level is rigid enough to be engaged by a series of asbestos faced rollers that continue to draw the ribbon of glass up a tower some 10 m high, after which the ribbon is cut into sheets and washed in a dilute acid to remove surface deposits. In the second method the glass is initially drawn in the vertical plane but it is turned over a roller so that it is drawn in the horizontal direction for some 60 m and passes into an annealing furnace, at the cold end of which it is cut into sheets.

Clear sheet glass is a transparent glass with 85% light transmission, with a fire finished surface, but because the two surfaces are never perfectly flat or parallel there is always some distortion of vision and reflection. BS 952 recommends three qualities for sheet glass:

1. Ordinary glazing quality (OQ) to be used for general glazing purposes.
2. Selected glazing quality (SGQ) for glazing work requiring a sheet glass above the ordinary glazing quality.
3. Special selected quality (SSQ) for high grade work where a superfine sheet glass is required such as cabinets.

Generally six thicknesses are produced ranging from 2-6 mm thick, the 2 mm thickness not being recommended for general glazing.

FLOAT GLASS

This is a transparent glass giving 85% light transmission and is a truly flat glass with undistorted vision. It is formed by floating a continuous ribbon of molten glass over a bath of liquid metal at a controlled rate and temperature. A general glazing quality and a selected quality are produced in six thicknesses ranging from 3-12 mm thick.

ROLLED AND ROUGH CAST GLASS

This is a term applied to a flat glass produced by a rolling process. Generally the glass produced in this manner is translucent which transmits light with varying degrees of diffusion so that vision is not clear. A wired transparent glass with 80% light transmission is, however, produced generally in one thickness of 6 mm. The glass is made translucent by rolling on to one face a texture or pattern which will give 70-85% light transmission. Rough cast glass has an irregular texture to one side; wired rough cast glass comes in two forms, Georgian wired, which has a 12 mm square mesh electrically welded wire reinforcement, or hexagonally wired which is reinforced with hexagonal wire of approximately 20 mm mesh.

Rough cast glass is produced in 5, 6 and 10 mm thicknesses and is made for safety and fire resistant glazing purposes.

Glazing

The securing of glass in prepared openings such as in doors, windows and partitions is termed glazing. The actual calculations for glass sizing are beyond the scope of this book but it is logical that as the area of glass pane increases so must its thickness also increase; similarly position, wind load and building usage must be taken into account.

GLAZING WITHOUT BEADS

This is a suitable method for general domestic window and door panes, the glass is bedded in a compound and secured with sprigs, pegs or clips and fronted with a weathered surface putty. Putty is a glazing compound which will require a protective coating of paint as soon as practicable after glazing. Two kinds of putty are in general use:

1. **Linseed oil putty**: for use with primed wood members and is made from linseed oil and whiting, usually to the recommendations of BS 544.
2. **Metal casement putty**: for use with metal or non-absorbent wood members and is made from refined vegetable drying oils and finely ground chalk.

The glass pane should be cut to allow a minimum clearance of 2 mm all round for both wood and metal frames. Sufficient putty is applied to the rebate to give at least 2 mm of back putty when the glass is pressed into the rebate, any surplus putty being stripped off level or at an angle above the rebate. The glass should be secured with sprigs or clips at not more than 440 mm centres and finished off on the front edge with a weathered putty fillet so that the top edge of the fillet is at or just below the sight line (see Fig. III.15).

GLAZING WITH BEADS

For domestic work glazing with beads is generally applied to good class joinery. The beads should be secured with either panel pins or screws—for hardwoods it is usual to use cups and screws. The glass is bedded in a compound or a suitable glazing felt mainly to prevent damage by vibration to the glass. Beads are usually mitred at the corners to give continuity of any moulding. Beads for metal windows are usually supplied with the surround or frame and fixing of glass should follow the manufacturer's instructions (se Fig. III.15).

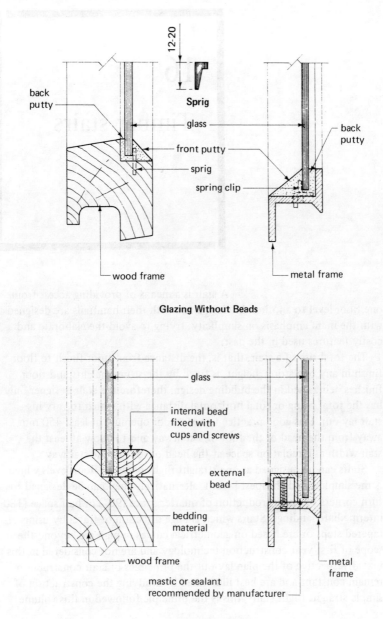

12-20

Sprig

back putty

glass

front putty

sprig

back putty

spring clip

wood frame

metal frame

Glazing Without Beads

glass

internal bead fixed with cups and screws

external bead

bedding material

wood frame

metal frame

mastic or sealant recommended by manufacturer

Glazing With Beads

Fig. III.15 Glazing details

16
Timber stairs

A stair is a means of providing access from one floor level to another. Modern stairs with their handrails are designed with the main emphasis on simplicity, trying to avoid the elaborate and costly features used in the past.

The total rise of a stair, that is, the distance from floor finish to floor finish in any one storey height, is fixed by the storey heights and floor finishes being used in the building design; therefore the stair designer only has the total going or total horizontal distance with which to vary his stair lay-out. It is good practice to keep door openings at least 450 mm away from the head or the foot of a stairway and to allow at least the stair width as circulation space at the head or foot of the stairway.

Stairs can be designed as one straight flight between floor levels which is the simplest and cheapest layout, alternatively they can be designed to turn corners by the introduction of quarter space (90°) or half space (180°) intermediate landings. Stairs which change direction of travel by using tapered steps or are based on geometrical curves in plan are beyond the scope of first year construction technology and are not considered in this text. Irrespective of the plan lay-out the principles of stair construction remain constant and are best illustrated by studying the construction of simple straight flight stairs; this is the principle followed in this volume.

Stair terminology
Stairwell: the space in which the stairs and landings are housed.

Stairs: the actual means of ascension or descension from one level to another.

Tread: the upper surface of a step on which the foot is placed.

Nosing: the exposed edge of a tread. usually projecting with a square. rounded or splayed edge.

Riser: the vertical member between two consecutive treads.

Step: riser plus tread.

Going: the horizontal distance between two consecutive risers or, as defined in Approved Document K, the distance measured on plan between two consecutive nosings.

Rise: the vertical height between two consecutive treads.

Flight: a series of steps without a landing.

Newel: post forming the junction of flights of stairs with landings or carrying the lower end of strings.

Strings: the members receiving the ends of steps which are generally housed to the string and secured by wedges, called wall or outer strings according to their position.

Handrail: protecting member usually parallel to the string and spanning between newels. This could be attached to a wall above and parallel to a wall string.

Baluster: the vertical infill member between a string and handrail.

Pitch line: a line connecting the nosings of all treads in any one flight.

BUILDING REGULATIONS PART K

These studies cover only those stairs intended for use in connection with dwellings—other forms of stairways are covered by *Construction Technology*, Volumes 2 and 3.

Approved Document K defines two forms of stairs:

1. **Common stairway:** which is an internal or external stairway which forms part of a building and is intended for common use in connection with two or more dwellings.

2. **Private stairway:** which is an internal or external stairway of steps with straight nosings on plan which forms part of a building and is either within a dwelling or intended for use solely in connection with one dwelling.

Construction and design

It is essential to keep the dimensions of the treads and risers constant throughout any flight of steps to reduce the risk of accidents by changing the rhythm of movement up or down the stairway. The height of the individual step rise is calculated by dividing the total rise by the chosen number of risers. The individual step going is chosen to suit the floor area available so that it, together with the rise, meets the requirements of the Building Regulations (see Fig. III.16). It is important to note that in any one flight there will be one more riser than treads since the last tread is in fact the landing.

Stairs are constructed by joining the steps into the spanning members or strings by using housing joints, glueing and wedging the steps into position to form a complete and rigid unit. Small angle blocks can be glued at the junction of tread and riser in a step to reduce the risk of slight movement giving rise to the annoyance of creaking. The flight can be given extra rigidity by using triangular brackets placed under the steps on the centre line of the flight. The use of a central beam or carriage piece with rough brackets as a support is only used on wide stairs over 1 200 mm, especially where they are intended for use as a common stairway (see Fig. III.17).

Stairs can be designed to be either fixed to a wall with one outer string, fixed between walls or freestanding—the majority have one wall string and one outer string. The wall string is fixed directly to the wall along its entire length or is fixed to timber battens plugged to the wall, the top of the string being cut and hooked over the trimming member of the stairwell. The outer string is supported at both ends by a newel post which in the case of the bottom newel rests on the floor; and, in the case of the upper newel, it is notched over and fixed to the stairwell trimming member. If the upper newel is extended to the ground floor to give extra support, it is called a storey newel post. The newel posts also serve as the termination point for handrails which span between them and is then infilled with balusters, balustrade or a solid panel to complete the protection to the sides of the stairway (see Fig. III.18).

If the headroom distance is critical it is possible to construct a bulkhead arrangement over the stairs as shown in Fig. III.19; this may give the increase in headroom required to comply with the Building Regulations. The raised floor in the room over the stairs can be used as a shelf or form the floor of a hanging cupboard.

164

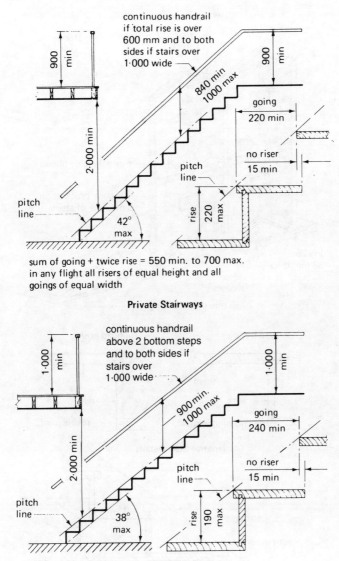

continuous handrail if total rise is over 600 mm and to both sides if stairs over 1·000 wide

900 min

900 min

840 min 1000 max

2·000 min

going 220 min

no riser 15 min

pitch line

pitch line

rise 220 max

42° max

sum of going + twice rise = 550 min. to 700 max. in any flight all risers of equal height and all goings of equal width

Private Stairways

continuous handrail above 2 bottom steps and to both sides if stairs over 1·000 wide

1·000 min

1·000 min

900 min. 1000 max

2·000 min

going 240 min

no riser 15 min

pitch line

pitch line

rise 190 max

38° max

sum of going + twice rise = 550 min. to 700 max. in any flight all risers of equal height and all goings of equal width.

Common Stairways

Fig. III.16 Timber stairs and Approved Document K

165

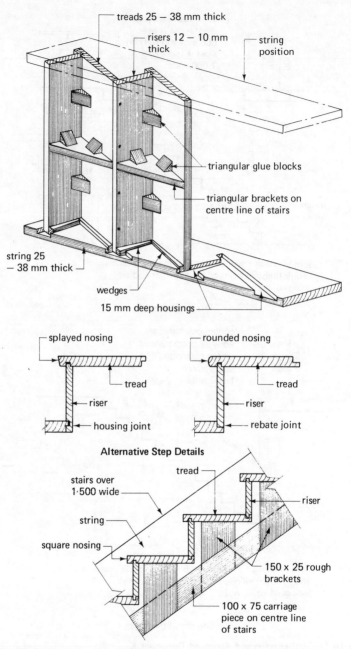

treads 25 – 38 mm thick

risers 12 – 10 mm thick

string position

triangular glue blocks

triangular brackets on centre line of stairs

string 25 – 38 mm thick

wedges

15 mm deep housings

splayed nosing

tread

riser

housing joint

rounded nosing

tread

riser

rebate joint

Alternative Step Details

stairs over 1·500 wide

tread

string

square nosing

riser

150 × 25 rough brackets

100 × 75 carriage piece on centre line of stairs

Fig. III.17 Stair construction details

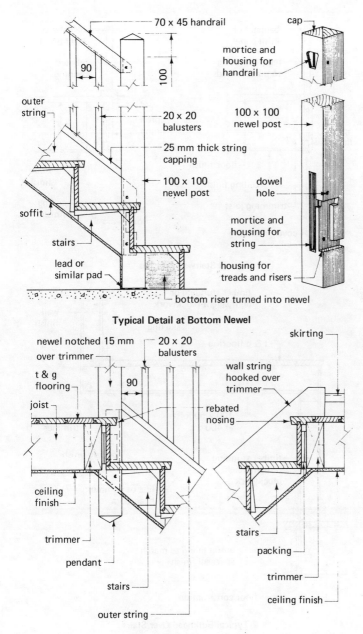

70 x 45 handrail

cap

90

100

mortice and housing for handrail

outer string

20 x 20 balusters

100 x 100 newel post

25 mm thick string capping

100 x 100 newel post

dowel hole

soffit

mortice and housing for string

stairs

lead or similar pad

housing for treads and risers

bottom riser turned into newel

Typical Detail at Bottom Newel

newel notched 15 mm over trimmer

20 x 20 balusters

skirting

t & g flooring

90

wall string hooked over trimmer

joist

rebated nosing

ceiling finish

trimmer

pendant

stairs

packing

stairs

trimmer

outer string

ceiling finish

Typical Details at Landing

Fig. III.18 Stair support and fixing details

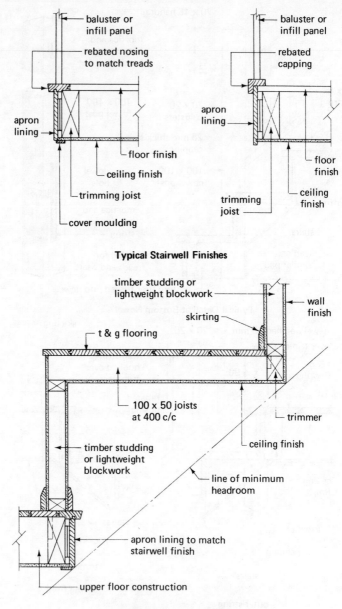

Typical Stairwell Finishes

Typical Bulkhead Over Stairs

Fig. III.19 Stairwell finishes and bulkhead details

168

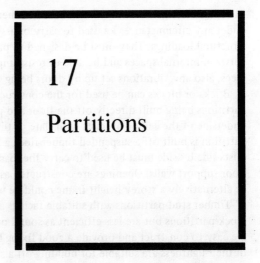

17
Partitions

Internal walls which divide the interior of a building into areas of accommodation and circulation are called partitions and these can be classified as load bearing or non-load bearing partitions.

Load bearing partitions

These are designed and constructed to receive superimposed loadings and transmit these loads to a foundation (see Fig. II.33). Generally load bearing partitions are constructed of bricks or blocks bonded to the external walls (see Figs. II.12-II.13, II.14, II.19 and II.20). Openings are formed in the same manner as for external walls; a lintel spans the opening carrying the load above to the reveals on either side of the opening. Fixings can usually be made direct into block walls, whereas walls of bricks require to be drilled and plugged to receive nails or screws. Apart from receiving direct fixings block partitions are lighter, cheaper, quicker to build and have better thermal insulation properties than brick walls but their sound insulation values are slightly lower. For these reasons blockwork is usually specified for load bearing partitions in domestic work. Load bearing partitions are, because of their method of construciton, considered to be permanently positioned.

Non-load bearing partitions

These partitions, like load bearing partitions, must be designed and constructed to carry their own weight and any

169

fittings or fixings which may be attached to them, but they must not under any circumstances be used to carry or assist in the transmission of structural loadings. They must be designed to provide a suitable division between internal spaces and be able to resist impact loadings on their faces, also any vibrations set up by doors being closed or slammed.

Bricks or blocks can be used for the construction of non-load bearing partitions being built directly off the floor and pinned (or wedged) to the underside of the ceiling or joists with slate or tile slips and mortar. If the partition is built off a suspended timber floor a larger joist section or two joists side by side must be used to carry the load of the partition to the floor support walls. Openings are constructed as for load bearing walls or alternatively a storey height frame could be used.

Timber stud partitions with suitable facings are lighter than brick or block partitions but are less efficient as sound or thermal insulators. They are easy to construct and provide a good fixing background and because of their lightness are suitable for building off a suspended timber floor. The basic principle is to construct a simple framed grid of timber to which a dry lining such as plywood, plasterboard or hardboard can be attached. The lining material will determine the spacing of the uprights or studs to save undue wastage in cutting the boards to terminate on the centre line of a stud. To achieve a good finish it is advisable to use studs which have equal thicknesses since the thin lining materials will follow any irregularity of the face of the studwork. Openings are formed by framing a head between two studs and fixing a lining or door frame into the opening in the stud partition; typical details are shown in Fig. III.20.

Plasterboard bonded on either side of a strong paper cellure core to form rigid panels are suitable as non-load bearing partitions. These are fixed to wall and ceiling battens and supported on a timber sole plate. Timber blocks are inserted into the core to provide the fixing medium for door frames or linings and skirtings (see Fig. III.21). This form of partition is available with facings suitable for direct decoration, plaster coat finish or with a plastic face as a self finish.

Compressed strawboard panels provide another alternative method for non-load bearing partitions. The storey height panels are secured to a sole plate and a head plate by skew or tosh nailing through the leading edge. The 3 mm joint between consecutive panels is made with an adhesive supplied by the manufacturer. Openings are formed by using storey height frames fixed with 100 mm screws direct into the edge of the panel. The joints are covered with a strip of hessian scrim and the whole partition is given a skim coat of board plaster finish (see Fig. III.21).

Partially prefabricated partitions, such as the plasterboard and straw-board described above, can be erected on site without undue mess. Being mainly a dry construction they reduce the drying time required with the

170

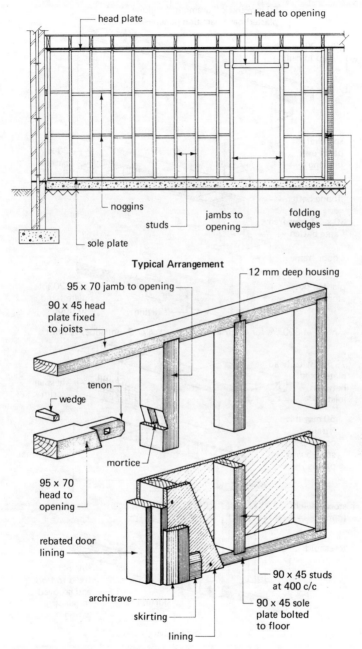

head plate

head to opening

noggins

studs

jambs to opening

folding wedges

sole plate

Typical Arrangement

95 x 70 jamb to opening

12 mm deep housing

90 x 45 head plate fixed to joists

tenon

wedge

mortice

95 x 70 head to opening

rebated door lining

architrave

skirting

lining

90 x 45 studs at 400 c/c

90 x 45 sole plate bolted to floor

Fig. III.20 Typical timber stud partition

171

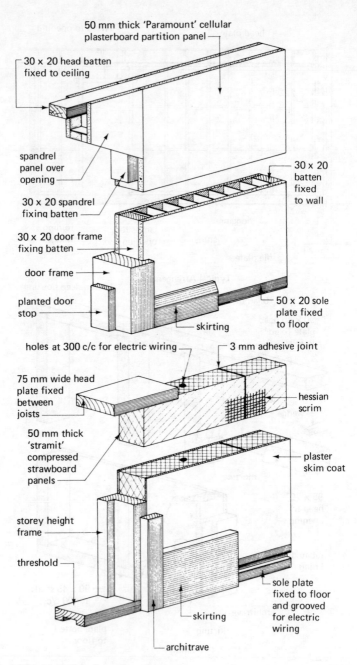

50 mm thick 'Paramount' cellular plasterboard partition panel

30 x 20 head batten fixed to ceiling

spandrel panel over opening

30 x 20 spandrel fixing batten

30 x 20 door frame fixing batten

door frame

planted door stop

30 x 20 batten fixed to wall

50 x 20 sole plate fixed to floor

skirting

holes at 300 c/c for electric wiring

3 mm adhesive joint

75 mm wide head plate fixed between joists

hessian scrim

50 mm thick 'stramit' compressed strawboard panels

plaster skim coat

storey height frame

threshold

architrave

skirting

sole plate fixed to floor and grooved for electric wiring

Fig. III.21 Typical preformed partitions

172

traditional brick and block walls. They also have the advantage, like timber stud partitions, of being capable of removal or repositioning without causing serious damage to the structure or without causing serious problems to a contractor.

18

Finishes–floor, wall and ceiling

FLOOR FINISHES

The type of floor finish to be applied to a floor will depend upon a number of factors such as type of base, room usage, degree of comfort required, maintenance problems, cost, appearance, safety and individual preference.

Floor finishes can be considered under three main headings:

1. *In situ* **floor finishes:** the finishes which are mixed on site, laid in a fluid state, allowed to dry and set to form a hard jointless surface.
2. **Applied floor finishes:** those finishes which are supplied in tile or sheet form and are laid on to a suitably prepared base.
3. **Timber floor finishes:** these are boards, sheets and blocks of timber laid on or attached to a suitable structural frame or base.

In situ floor finishes

MASTIC ASPHALT

This is a naturally occurring bituminous material obtained from asphalt lakes like those in Trinidad or it can also be derived from crude oil residues. Trinidad lake asphalt is used as a matrix or cement to a suitably graded mineral aggregate to form mastic asphalt as a material suitable for floor finishing. When laid mastic asphalt is impervious to water and is ideal for situations such as sculleries and wash rooms. It also forms a very good surface on which to apply thin tile and sheet finishes (for example,

PVC) and will at the same time fulfil the function of a damp-proof membrane.

Mastic asphalt is a thermoplastic material and has to be melted before it can be applied to form a jointless floor finish. Hot mastic asphalt is applied by means of a float at a temperature of between 180 and 210°C in a single 13 mm coat as a base for applied finishes or in a 16 mm single coat for a self finish. Any sound and rigid subfloor is suitable but a layer of ordinary black sheathing felt should be included between the subfloor and mastic asphalt to overcome the problems caused by differential movement. The finish obtained is smooth and hard but the colour range is limited to dark colours such as red, brown and black. A matt surface can be produced by giving the top surface a dusting of sand or powdered stone.

PITCH MASTIC

Pitch mastic is a similar material to mastic asphalt but is produced from a mixture of calcareous and/or siliceous aggregates bonded with coal tar pitch. It is laid to a similar thickness and in a similar manner to mastic asphalt with a polished or matt finish. Pitch mastic floors are resistant to water but have a better resistance to oil and fats than mastic asphalt and are therefore suitable for sculleries, washrooms and kitchens.

GRANOLITHIC

This is a mixture of Portland cement and granite chippings which can be applied to a 'green' concrete subfloor or to a cured concrete subfloor. Green concrete is a term used to describe newly laid concrete which is not more than three hours old. A typical mix for granolithic is 1 part cement : 1 part sand : 2 parts granite chippings (5-10 mm free from dust) by volume. The finish obtained is very hard wearing, noisy and cold to touch; it is used mainly in situations where easy maintenance and durability are paramount such as a common entrance hall to a block of flats.

If granolithic is being applied to a green concrete subfloor as a topping it is applied in a single layer approximately 20 mm thick in bay sizes not exceeding 28 m^2 and trowelled to a smooth surface. This method will result in a monolithic floor and finish construction.

The surface of mature concrete will need to be prepared by hacking the entire area and brushing well to remove all the laitance before laying a single layer of granolithic which should be at least 40 mm thick. The finish should be laid, on a wet cement slurry coating to improve the bond, in bay sizes not exceeding 14 m^2.

175

MAGNESIUM OXYCHLORIDE

This is a composition flooring which is sometimes used as a substitute for asphalt since it has similar wearing and appearance properties. It is mixed on site using a solution of magnesium chloride with burned magnesite and fillers such as wood flour, sawdust, powdered asbestos or limestone. It is essential that the ingredients are thoroughly mixed in a dry state before the correct amount of solution is added. The mixed material is laid in one or two layers giving a total thickness of approximately 20 mm. The subfloor must be absolutely dry otherwise the finish will lift and crack, therefore a damp-proof membrane in the subfloor is essential.

Applied floor finishes

Many of the applied floor finishes are thin flexible materials and should be laid on a subfloor with a smooth finish. This is often achieved by laying a cement/sand bed or screed with a steel float finish to the concrete subfloor. The usual screed mix is 1:3 cement/ sand and great care must be taken to ensure that a good bond is obtained with the subfloor. If the screed is laid on green concrete a thickness of 12 mm is usually specified whereas screeds laid on mature concrete require a thickness of 40 mm. A mature concrete subfloor must be clean, free from dust and dampened with water to reduce the suction before applying the bonding agent to receive the screed. To reduce the possibility of drying shrinkage cracks screeds should not be laid in bays exceeding 15 m^2 in area.

FLEXIBLE PVC TILES AND SHEET

Flexible PVC is a popular hardwearing floor finish produced by a mixture of polyvinyl chloride resin, pigments and mineral fillers. It is produced as 300 x 300 mm square tiles or in sheet form up to 2 400 mm wide with a range of thicknesses from 1·5-3 mm. The floor tiles and sheet are fixed with an adhesive recommended by the manufacturer and produce a surface suitable for most situations. PVC tiles, like all other small unit coverings, should be laid from the centre of the area towards the edges, so that if the area is not an exact tile module an even border of cut tiles is obtained.

THERMOPLASTIC TILES

These are sometimes called asphalt tiles and are produced from coumarone indene resins, fillers and pigments; the earlier use of asphalt binders limited the range of colours available. These tiles are hardwearing, moisture resistant and are suitable for most situations, being produced generally as a 225 mm square tile either 3 or 4·5 mm thick. To make them pliable they are usually heated before being fixed with a bituminous adhesive.

RUBBER TILES AND SHEET

Solid rubber tiles or sheet is produced from natural or synthetic rubber compounded with fillers to give colour and texture, the rubber content being not less than 35%. The covering is hardwearing, quiet and water resistant and suitable for bathrooms and washrooms. Thicknesses range from 3-6·5 mm with square tile sizes ranging from 150-1 200 mm; sheet widths range from 900-1 800 mm. Fixing is by rubber based adhesives, as recommended by the manufacturer, to a smooth surface.

LINOLEUM

Linoleum is produced in sheet or tile form from a mixture of drying oils, resins, fillers and pigments which is pressed on to a hessian or bitumen saturated felt paper backing. Good quality linoleum has the pattern inlaid or continuing through the thickness whereas the cheaper quality has only a printed surface pattern. Linoleum gives a quiet, resilient and hardwearing surface suitable for most domestic floors. Thicknesses vary from 2-6·5 mm for a standard sheet width of 1 800 mm; tiles are usually 300 mm square with the same range of thicknesses. Fixing of linoleum tiles and sheet is by adhesive to any dry smooth surface, although the adhesive is sometimes omitted with the thicker sheets.

CARPET

The chief materials used in the production of carpets are nylon, acrylics and wool or mixtures of these materials. There is a vast range of styles, types, patterns, colours, qualities and sizes available for general domestic use in dry situations since the resistance of carpets to dampness is generally poor. To obtain maximum life carpets should be laid over an underlay of felt or latex and secured by adhesives, nailing around the perimeter or being stretched over and attached to special edge fixing strips. Carpet is supplied in narrow or wide rolls, carpet squares (600 x 600 x 25 mm thick) are also available for covering floors without the use of adhesives; these rely on the interlocking of the edge fibres to form a continuous covering.

CORK TILES AND CARPET

Cork tiles are cut from baked blocks of granulated cork where the natural resins act as the binder. The tiles are generally 300 mm square with thicknesses of 5 mm upwards according to the wearing quality required and are supplied in three natural shades. They are hardwearing, quiet and resilient but unless treated with a surface sealant they may collect dirt and grit. Fixing is by using an adhesive recommended by the manufacturer, with the addition of nine steel pins around the edge to prevent curling.

Cork carpet is a similar material but it is made pliable by bonding the cork granules with linseed oil and resins on to a jute canvas backing. It is laid in the same manner as described above for linoleum and should be treated with a surface sealant to resist dirt and grit penetration.

QUARRY TILES

The term 'quarry' is derived from the Italian word *quadro* meaning square and does not mean that the tiles are cut or won from an open excavation or quarry. They are made from ordinary or unrefined clays worked into a plastic form, pressed into shape and hard burnt. Being hardwearing and with a good resistance to water they are suitable for kitchens and entrance halls but they tend to be noisy and cold. Quarry tiles are produced as square tiles in sizes ranging from 100 x 100 x 20 mm to 225 x 225 x 32 mm thick.

Three methods of laying quarry tiles are recommended to allow for differential movement due to drying shrinkage or thermal movement of the screed. The first method is to bed the tiles in a 10-15 mm thick bed of cement mortar over a separating layer of sheet material such as polythene, which could perform the dual function of damp-proof membrane and separating layer. To avoid the use of a separating layer the 'thick bed' method can be used. With this method the concrete subfloor should be dampened to reduce suction and then covered with a semi-dry 1 : 4 cement/sand mix to a thickness of approximately 40 mm. The top surface of the compacted semi-dry 'thick bed' should be treated with a 1 : 1 cement/sand grout before tapping the tiles in to the grout with a wood beater. Alternatively a dry cement may be trowelled into the 'thick bed' surface before bedding in the tiles. Special cement based adhesives are also available for bedding tiles to a screed and these should be used in accordance with the manufacturer's instructions to provide a thin bed fixing.

Clay tiles may expand probably due to physical adsorption of water and chemical hydration and for this reason an expansion joint of compressible material should be incorporated around the perimeter of the floor (see Fig. III.22). In no circumstances should the length or width of a quarry tile floor exceed 7 500 mm without an expansion joint. The joints between quarry tiles are usually grouted with a 1 : 1 cement/sand grout and then, after cleaning, the floor should be protected with sand or saw-dust for four or five days. There is a wide variety of tread patterns available to provide a non-slip surface, also a wide range of fittings are available to form skirtings and edge coves (see Fig. III.22).

PLAIN CLAY FLOOR TILES

These are similar to quarry tiles but are produced from refined natural clays which are pressed after grinding and tempering into the desired shape before being fired at a high temperature. Plain clay floor tiles, being denser than quarry tiles, are made as smaller and thinner units ranging from 75 x 75 mm to 150 x 150 mm, all sizes being 13 mm thick.

Laying, finishes and fittings available are all as described for quarry tiles.

Timber floor finishes

Timber is a very popular floor finish with both designer and user because of its natural appearance, resilience and warmth. It is available as a board, strip, sheet or block finish and if attached to joists, as in the case of a suspended timber floor, it also acts as the structural decking.

TIMBER BOARDS

Softwood timber floor boards are joined together by tongued and grooved joints along their edges and are fixed by nailing to the support joists or fillets attached to a solid floor. The boards are butt jointed in their length, the joints being positioned centrally over the supports and staggered so that butt joints do not occur in the same position in consecutive lengths. The support spacing will be governed by the spanning properties of the board (which is controlled by its thickness) and supports placed at 400 mm centres are usual for boards of 19 and 22 mm thick. The tongue is positioned slightly off centre and the boards should be laid with the narrow shoulder on the underside to give maximum wear. It is essential that the boards are well cramped together before being fixed to form a tight joint and that they are laid in a position where they will not be affected by dampness. Timber is a hygroscopic material and will therefore swell and shrink as its moisture content varies—ideally this should be maintained at around 12%.

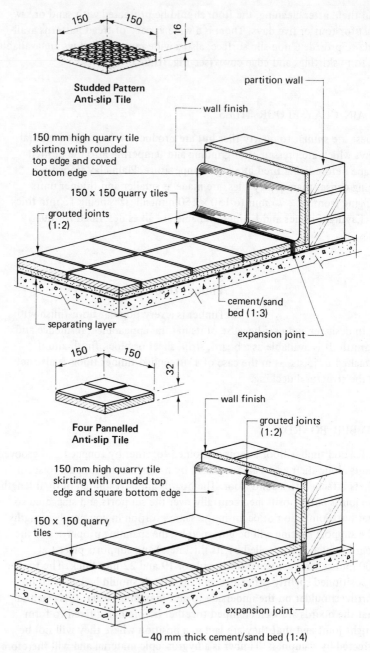

150
150
16

**Studded Pattern
Anti-slip Tile**

partition wall

wall finish

150 mm high quarry tile
skirting with rounded
top edge and coved
bottom edge

150 x 150 quarry tiles

grouted joints
(1:2)

cement/sand
bed (1:3)

separating layer

expansion joint

150
150
32

**Four Pannelled
Anti-slip Tile**

wall finish

grouted joints
(1:2)

150 mm high quarry tile
skirting with rounded top
edge and square bottom edge

150 x 150 quarry
tiles

expansion joint

40 mm thick cement/sand bed (1:4)

Fig. III.22 Typical quarry tile floors

TIMBER STRIP

These are narrow boards being under 100 mm wide to reduce the amount of shrinkage and consequent opening of the joints. Timber strip can be supplied in softwood or hardwood and is considered to be a superior floor finish to boards. Jointing and laying is as described for boards, except that hardwood strip is very often laid one strip at a time and secret nailed (see Fig. III.23).

TIMBER SHEET FLOOR FINISH

Chipboard is manufactured from wood chips or shavings bonded together with thermosetting synthetic resins, and forms rigid sheets 12·5 and 19 mm thick which is suitable as a floor finish. The sheets are fixed by nailing or screwing to support joists or fillets and should have a plywood reinforcement in the form of a loose tongue to all adjoining edges which need to be grooved to receive the tongue. If the sheet is used as an exposed finish a coating of sealer should be used. Alternatively chipboard can be used as a decking to which a thin tile or carpet finish can be applied.

Tongued and grooved boards of 600 mm width are also available as a floor decking material which do not need to be jointed over a joist.

WOOD BLOCKS

These are small blocks of timber usually of hardwood which are designed to be laid in set patterns. Lengths range from 150-300 mm with widths up to 89 mm; the width being proportional to the length to enable the various patterns to be created. Block thicknesses range from 20-30 mm thick and the final thickness after sanding and polishing being about 5-10 mm thinner.

The blocks are jointed along their edges with a tongued and grooved joint and have a rebate, chamfer or dovetail along the bottom longitudinal edges to take up any surplus adhesive used for fixing. Two methods can be used for fixing wood blocks: the first uses hot bitumen and the second a cold latex bitumen emulsion. If hot bitumen is used the upper surface of the subfloor is first primed with black varnish to improve adhesion, and then, before laying, the bottom face of the block is dipped into the hot bitumen. The cold adhesive does not require a priming coat to the subfloor. Blocks, like tiled floors, should be laid from the centre of the floor towards the perimeter, which is generally terminated with a margin border. To allow for moisture movement a cork expansion strip should be placed around the entire edge of the block floor (see Fig. III.23).

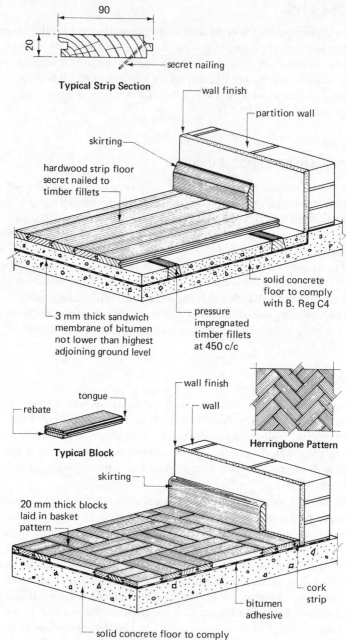

90

20

secret nailing

Typical Strip Section

wall finish

partition wall

skirting

hardwood strip floor
secret nailed to
timber fillets

solid concrete
floor to comply
with B. Reg C4

3 mm thick sandwich
membrane of bitumen
not lower than highest
adjoining ground level

pressure
impregnated
timber fillets
at 450 c/c

tongue

rebate

Typical Block

wall finish

wall

Herringbone Pattern

skirting

20 mm thick blocks
laid in basket
pattern

cork
strip

bitumen
adhesive

solid concrete floor to comply
with B. Reg. C4 with floated finish

Fig. III.23 Typical timber floor finishes

PARQUET

This is a superior form of wood block flooring made from specially
selected hardwoods chosen mainly for their decorative appearance.
Parquet blocks are generally smaller and thinner than hardwood blocks
and are usually fixed to a timber subfloor which is level and smooth.
Fixing can be by adhesives or secret nailed, alternatively they can be
supplied as a patterned panel fixed to a suitable backing sheet in panel
sizes from 300-600 mm square.

Wall finishes

External brickwork with an exposed face of
facing bricks is a self finish and requires no further treatment. External
walls of common bricks or blocks can be treated to give an acceptable
appearance by the application of paint, an applied wall finish such as
rendering or be clad with boards or tiles. Internal walls or partitions can
be built with a fair face of natural materials such as bricks or stone but
generally it is cheaper to use a material such as blocks with an applied
finish like plaster, dry lining or glazed tiles.

External rendering

This is a form of plastering using a mixture
of cement and sand, or cement, lime and sand, applied to the face of a
building to give extra protection against the penetration of moisture or to
provide a desired texture. It can also be used in the dual capacity of
providing protection and appearance.

The rendering must have the properties of durability, moisture
resistance and an acceptable appearance. The factors to be taken into
account in achieving the above requirements are mix design, bond to the
backing material, texture of surface, degree of exposure of the building and
the standard of workmanship in applying the rendering.

Cement and sand mixes will produce a strong moisture resistant rendering but one which is subject to cracking due to high drying shrinkage.
These mixes are used mainly on members which may be vulnerable to
impact damage such as columns. Cement, lime and sand mixes have a lower
drying shrinkage but are more absorbent than cement and sand mixes;
they will, however, dry out rapidly after periods of rain and are therefore
the mix recommended for general use.

The two factors which govern the proportions to be used in a mix are:

1. Background to which the rendering is to be applied.
2. Degree of exposure of the building.

The two common volume mix ratios are:

(a) $1 : \frac{1}{2} : 4\text{-}4\frac{1}{2}$ cement : lime : sand which is used for dense, strong backgrounds of moderate to severe exposure and for application to metal lathing or expanded metal backgrounds.

(b) $1 : 1 : 5\text{-}6$ cement : lime : sand which is for general use.

If the rendering is to be applied in a cold, damp situation the cement content of the mix ratio should be increased by 1. The final coat is usually a weaker mix than the undercoats, for example, when using an undercoat mix $1 : 1 : 6$ the final coat mix ratio should be $1 : 2 : 9$, this will reduce the amount of shrinkage in the final coat. The number of coats required will depend upon the surface condition of the background and the degree of exposure. Generally a two coat application is acceptable, except where the background is very irregular or the building is in a position of severe exposure when a three coat application would be specified. The thickness of any one coat should not exceed 15 mm and each subsequent coat thickness is reduced by approximately 3 mm to give a final coat thickness of 6-10 mm.

Various textured surfaces can be obtained on renderings by surface treatments such as scraping the surface with combs, saw blades or similar tools to remove a surface skin of mortar. These operations are carried out some three to four hours after the initial application of the rendering and before the final set takes place.

Alternative treatments are:

1. **Roughcast:** a wet plaster mix of 1 part cement : $\frac{1}{2}$ part lime : $1\frac{1}{2}$ parts shingle : 3 parts sand which is thrown on to a porous coat of rendering to give an even distribution.
2. **Pebbledash:** selected aggregate such as pea shingle is dashed or thrown on to a rendering background before it has set and is tamped into the surface with a wood float to obtain a good bond.
3. **Spattered finishes:** these are finishes applied by a machine (which can be hand operated), guns or sprays using special mixes prepared by the machine manufacturers.

Plastering

Plastering like brickwork is one of the old established crafts in this country having been introduced by the Romans. The plaster used was a lime plaster which generally has been superseded by gypsum plasters. The disadvantages of lime plastering are:

1. Drying shrinkage which causes cracking.

2. Slow drying out process which can take several weeks causing delays for the following trades.
3. Need to apply lime plaster in several coats, usually termed render, float and set, to reduce the amount of shrinkage.

Any plaster finish must fulfil the functions of camouflaging irregularities in the backing wall, provide a smooth continuous surface which is suitable for direct decoration and be sufficiently hard to resist damage by impact upon its surface; gypsum plasters fulfil these requirements.

Gypsum is a crystalline combination of calcium sulphate and water. Deposits of suitable raw material are found in several parts of England and after crushing and screening the gypsum is heated to dehydrate the material. The amount of water remaining at the end of this process defines its class under BS 1191. If powdered gypsum is heated to about 170°C it loses about three-quarters of its combined water and is called hemi-hydrate gypsum plaster but is probably better known as 'plaster of Paris'. If a retarder is added to the hemi-hydrate plaster a new class of finishing plaster is formed to which the addition of expanded perlite and other additives will form a one-coat or universal plaster, a renovating grade plaster or a spray plaster for application by a spray machine.

Part I of BS 1911 lists two classes of plaster:

Class A: hemi-hydrate plaster of Paris.
Class B: retarded hemi-hydrate finishing plasters.

Part 2 of BS 1191 covers pre-mixed lightweight plasters which are defined as plasters consisting of suitable lightweight aggregates and retarded hemi-hydrate gypsum plasters complying with BS 1191, Part 1, Class B.

Class A or plaster of Paris is a hemi-hydrate plaster having approximately 25% of its original combined water left and has a very rapid setting time of about ten minutes. It can be used neat or mixed with a little sand and is suitable for use as a filler, moulding and repair work.

Class B is a hemi-hydrate plaster to which an animal protein such as keratin has been added during manufacture to retard the setting time.

Class A and B plasters are supplied in non-returnable multi-walled paper sacks of 50 kg capacity and care must be taken not to tear the sacks before the plaster is used. The absorption of moisture by gypsum plasters shortens the normal setting time of approximately one and a half hours which may reduce the strength of the set plaster. It follows therefore that all plasters should be stored in dry conditions preferably on timber palettes.

Gypsum plasters are not suitable for use in temperatures exceeding 43°C and should not be applied to frozen backgrounds. However, plasters can be applied under frosty conditions providing the surfaces are adequately protected from freezing after application. The dry bagged plaster in store is unaffected by low temperatures.

The plaster should be mixed in a clean plastic or rubber bucket using clean water only. Cleanliness is imperative since any set plaster left in the mixing bucket from a previous mixing will shorten the setting time which may reduce the strength of the plaster when set.

Premixed plasters are primarily Class B plasters incorporating lightweight aggregates such as expanded perlite and exfoliated vermiculite. Perlite is a glassy volcanic rock combining in its chemical composition a small percentage of water. When subjected to a high temperature the water turns into steam thus expanding the natural perlite to many times its original volume. Vermiculite is a form of mica with many ultra thin layers between which are minute amounts of water, which expand or exfoliate the natural material to many times its original volume when subjected to high temperatures by the water turning into steam and forcing the layers of flakes apart. The density of a lightweight plaster is about one-third that of a comparable sanded plaster and it has a thermal value of about three times that of sanded plasters, resulting in a reduction of heat loss, less condensation and a reduction in the risk of pattern staining. They also have superior adhesion properties to all backgrounds including smooth dry clean concrete.

The choice of plaster mix, type and number of coats will depend upon the background or surface to which the plaster is to be applied. Roughness and suction properties are two of the major considerations. The roughness can affect the keying of the plaster to its background, special keyed bricks are available for this purpose (see Fig. II.10); or alternatively the joints can be raked out to a depth of 15-20 mm as the wall is being built. *In situ* concrete can be cast using sawn formwork giving a rough texture hence forming a suitable key. Generally all lightweight concrete blocks provide a suitable key for the direct application of plasters. Bonding agents in the form of resin emulsions are available for smooth surfaces—these must be applied in strict accordance with the manufacturer's instructions to achieve satisfactory results.

The suction properties of the background can affect the drying rate of the plaster by absorbing the moisture of the mix: too much suction can result in the plaster failing to set properly—thus losing its adhesion to its background; too little suction can give rise to drying shrinkage cracks due to the retention of excess water in the plaster.

The actual mix chosen will depend upon the class of plaster used and the properties of the background material. Cement backing to plasters usually have a volume mix of 1 : 1 : 6 cement : lime : sand, whereas finishing plaster coats can be applied neat or with up to 25% slaked lime putty by volume,

Undercoat plasters are applied by means of a wooden float or rule worked between dots or runs of plaster to give a true and level surface. The runs or rules and dots are of the same mix as the cement backing coat and are positioned over the background at suitable intervals to an accurate level so that the main application of plaster can be worked around the guide points. The upper surface of the undercoat plaster should be scored or scratched to provide a suitable key for the finishing coat. The thin finishing coat of plaster is applied to a depth of approximately 3 mm and finished with a steel float to give a smooth surface.

Dry lining techniques

External walls or internal walls and partitions can be dry lined with a variety of materials which can be self finished, ready for direct decoration or have a surface suitable for a single final coat of board finish plaster. The main advantages of dry lining techniques are speed, reduction in the amount of water used in the construction of buildings thus reducing the drying out period and in some cases increased thermal insulation.

Suitable materials are hardboard, plywood, chipboard and plasterboard. Hardboard, plywood and chipboard are fixed to timber battens attached to the wall at centres to suit the spanning properties and module size of the board. Finishing can be a direct application of paint, varnish or wallpaper but masking the fixings and joints may present problems. As an alternative the joints can be made a feature of the design by the use of edge chamfers or by using moulded cover fillets.

Plasterboard consists of an aerated gypsum core encased in and bonded to specially prepared bonded paper liners. The grey coloured liner is intended for a skim coat of plaster and the ivory coloured liner is for direct decoration. Plasterboard can also be obtained with a veneer of polished aluminium foil to one side to act as a reflective insulator when the board is fixed adjacent to a cavity. Four types of board are produced: wallboard, baseboard, lath and plank, all of which can be used for the dry lining of walls.

Wallboard is manufactured to metric coordinated widths of 600, 900 and 1 200 mm with coordinated lengths from 1 800-3 000 mm and thicknesses of 9·5 and 12·7 mm. The board can be obtained with a tapered edge for a seamless joint, a square edge for cover fillet treatment or with a chamfered edge for a vee-joint feature.

Baseboard is produced with square edges as a suitable base for a single coat of board finish plaster. One standard width is manufactured with a thickness of 9·5 mm, lengths are 1 200, 1 219 and 1 372 mm. Before the plaster coat is applied the joints should be covered with a jute scrim 100 mm wide (see Fig. III.24).

Plasterboard lath is a narrow plasterboard with rounded edges which removes the need for a jute scrim over the joints. The standard width is 406 mm with similar lengths to baseboard and thicknesses of 9·5 and 12·7 mm.

The fixing of dry linings is usually by nails to timber battens suitably spaced, plumbed and levelled to overcome any irregularities in the background. Fixing battens are placed vertically between the horizontal battens fixed at the junctions of the ceiling and floor with the wall. It is advisable to treat all fixing battens with an insecticide and a fungicide to lessen the risk of beetle infestation and fungal attack. The spacing of the battens will be governed by the spanning properties of the lining material, fixing battens at 450 mm centres are required for 9·5 mm thick plasterboards and at 600 mm centres for 12·7 mm thick plasterboards; so placed that they coincide with the board joints.

Plasterboard can also be fixed to a solid background by the use of 'dots' or 'dabs'. The dots consist of a 75 x 50 mm thin pad of bitumen impregnated fibreboard secured to the background with board finish plaster. Dots are placed at 1 000 mm centres vertically and 1 800 mm centres horizontally and levelled in all directions over the entire wall surface. Using these initial dots as datum points, intermediate dots are placed so that they coincide with the board joints. Once the dots have set (about one hour being usual), thick dabs of board finish plaster are applied vertically between the dots so that they stand proud of the dots. The plasterboard is then placed in position and tapped firmly until it is in contact with the dots, a double headed nail is used to temporarily secure the boards to the dots whilst the dabs are setting, after which the nails are removed and the joints made good. It is recommended that tapered edge boards are always used for this method of fixing.

Glazed wall tiles

Two processes are used in the manufacture of these hard glazed tiles. Firstly, the body or 'biscuit' tile is made from materials such as china clay, ball clay, flint and limestone which are mixed by careful processes into a fine powder before being heavily pressed into the required shape and size. The tile is then fired at a temperature of up to 1 150°C. The second process is the glazing which is applied in the form of a mixed liquid consisting of fine particles of glaze and water. The coated tiles are then fired for a second time at a temperature of approximately 1 050°C when the glazing coating fuses to the surface of the tile. Gloss, matt and eggshell finishes are available together with a wide choice of colours, designs and patterns.

A range of fittings with rounded edges and mitres are produced to blend with the standard 150 x 150 mm square tiles, 5 or 6 mm thick, also a similar range of fittings are made for the 108 x 108 mm, 4 or 6 mm thick, square tiles. The appearance and easily cleaned surface of glazed tiles makes them suitable for the complete or partial tiling of bathrooms and the provision of splashbacks for sinks and basins.

Tiles are fixed with a suitable adhesive which can be of a thin bed of mastic adhesive or a bed of a cement based adhesive; the former requires a flat surface such as a mortar screed. Ceramic or glazed tiles are considered to be practically inert, therefore careful selection of the right adhesive to suit the backing and final condition is essential. Manufacturer's instructions or the recommendations of BS 5385 should always be carefully followed.

Glazed tiles can be cut easily using the same method employed for glass. The tile is scored on its upper face with a glass cutter, this is followed by tapping the back of the tile behind the scored line over a rigid and sharp angle such as a flat straight-edge.

CEILING FINISHES

Ceilings can be finished by any of the dry lining techniques previously described for walls. The usual method is a plasterboard base with a skim coat of plaster. The plasterboards are secured to the underside of the floor or ceiling joists with galvanised plasterboard nails to reduce the risk of corrosion to the fixings. If square edged plasterboards are used as the base a jute scrim over the joints is essential. The most vulnerable point in a ceiling to cracking is at the junction between the ceiling and wall, this junction should be strengthened with a jute scrim around the internal angle (see Fig. III.24), or alternatively

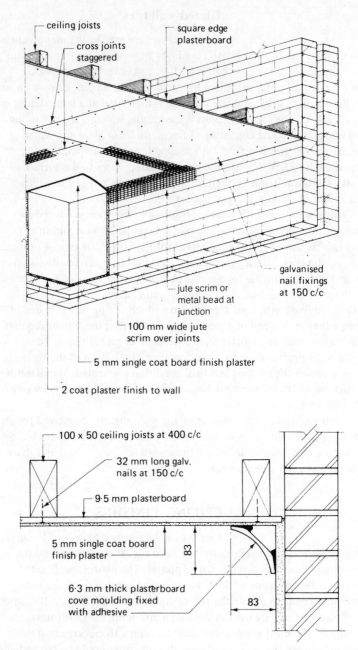

ceiling joists

cross joints
staggered

square edge
plasterboard

galvanised
nail fixings
at 150 c/c

jute scrim or
metal bead at
junction

100 mm wide jute
scrim over joints

5 mm single coat board finish plaster

2 coat plaster finish to wall

100 × 50 ceiling joists at 400 c/c

32 mm long galv.
nails at 150 c/c

9·5 mm plasterboard

5 mm single coat board
finish plaster

83

6·3 mm thick plasterboard
cove moulding fixed
with adhesive

83

Fig. III.24 Plasterboard ceilings

the junction can be masked with a decorative plasterboard or polystyrene cove moulding.

The cove moulding is made in a similar manner to plasterboard and is intended for direct decoration. Plasterboard cove moulding is jointed at internal and external angles with a mitred joint and with a butt joint in the running length. Any clean, dry and rigid background is suitable for the attachment of plasterboard cove which can be fixed in one of two ways. It can be secured by using a special water mixed adhesive applied to the contact edges of the moulding which is pressed into position; any surplus adhesive should be removed from the edges before it sets. Alternatively the cove moulding can be fixed with galvanised steel or brass screws to plugs or battens—fixings to the wall are spaced at 300 mm centres and to the ceiling at 600 mm centres. A typical plasterboard cove detail is shown in Fig. III.24.

Many forms of ceiling tiles are available for application to a joisted ceiling or solid ceiling with a sheet or solid background. Fixing to joists should be by concealed or secret nailing through the tongued and grooved joint. If the background is solid such as a concrete slab then dabs of a recommended adhesive are used to secure the tiles. Materials available include expanded polystyrene, mineral fibre, fibreboard and glass fibre with a rigid vinyl face.

Other forms of finish which may be applied to ceilings are sprayed plasters which can be of a thick or thin coat variety. Spray plasters are usually of a proprietary mixture applied by spraying apparatus directly on to the soffit giving a coarse texture which can be trowelled smooth if required. Various patterned ceiling papers are produced to give a textured finish. These papers are applied directly to the soffit or over a stout lining paper. Some ceiling papers are designed to be a self finish but others require one or more coats of emulsion paint.

19

Internal fixings and shelves

Internal fixings consist of trims in the form of skirtings, dado rails, frieze or picture rails, architraves and cornices; whereas fittings would include such things as cupboards and shelves.

SKIRTINGS

A skirting is a horizontal member fixed around the skirt or base of a wall primarily to mask the junction between a wall finish and a floor. It can be an integral part of the floor finish such as quarry tile or PVC skirtings or it can be made from timber, metal or plastic. Timber is the most common material used and is fixed by nails direct to the background but if it is a dense material that will not accept nails, plugs or special fixing bricks can be built into the wall. External angles in skirtings are formed by mitres but internal angles are usually scribed (see Fig. III.25).

ARCHITRAVES

These are mouldings cut and fixed around door and window openings to mask the joint between the wall finish and the frame. Like skirtings the usual material is timber but metal and plastic mouldings are available. Architraves are fixed with nails to the frame or lining and to the wall in a similar manner to skirtings if the architrave section is large (see Fig. III.25).

DADO RAILS

These are horizontal mouldings fixed in a position to prevent the walls

192

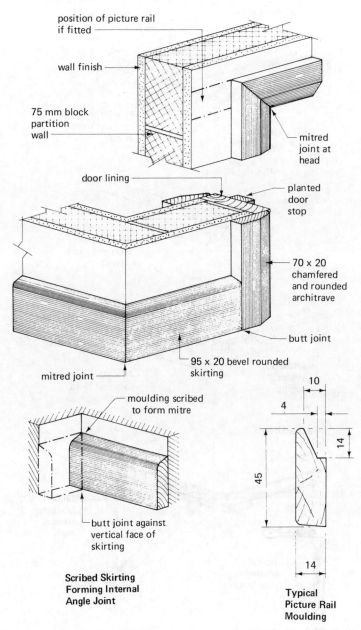

position of picture rail if fitted

wall finish

75 mm block partition wall

mitred joint at head

door lining

planted door stop

70 x 20 chamfered and rounded architrave

butt joint

95 x 20 bevel rounded skirting

mitred joint

moulding scribed to form mitre

butt joint against vertical face of skirting

Scribed Skirting Forming Internal Angle Joint

10

4

14

45

14

Typical Picture Rail Moulding

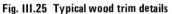

Fig. III.25 Typical wood trim details

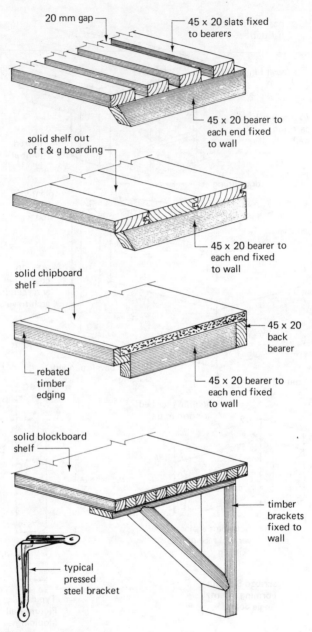

20 mm gap

45 x 20 slats fixed to bearers

45 x 20 bearer to each end fixed to wall

solid shelf out of t & g boarding

45 x 20 bearer to each end fixed to wall

solid chipboard shelf

45 x 20 back bearer

rebated timber edging

45 x 20 bearer to each end fixed to wall

solid blockboard shelf

timber brackets fixed to wall

typical pressed steel bracket

Fig. III.26 Shelves and supports

from being damaged by the backs of chairs pushed against them. They are very seldom used today since modern chair design renders them unnecessary. If used they are fixed by nails directly to the wall or to plugs inserted in the wall.

PICTURE RAILS

These are moulded rails fixed horizontally around the walls of a room from which pictures may be suspended and are usually positioned in line with the top edges of the door architrave. They can be of timber or metal and like the dado rail are very seldom used in modern domestic buildings. They would be fixed by nails in the same manner as dado rails and skirtings; a typical picture rail moulding is shown in Fig. III.25.

CORNICES

Cornices are timber or plaster ornate mouldings used to mask the junction between the wall and ceiling. These are very seldom used today having been superseded by the cove mouldings.

CUPBOARD FITTINGS

These are usually supplied as a complete fitting and only require positioning on site; they can be free standing or plugged and screwed to the wall. Built-in cupboards can be formed by placing a cupboard front in the form of a frame across a recess and then hanging suitable doors to the frame. Another method of forming built-in cupboards is to use a recessed partition wall to serve as cupboard walls and room divider, and attach to this partition suitable cupboard fronts and doors.

Shelving

Shelves can be part of a cupboard fitting or can be fixed to wall brackets or bearers which have been plugged and screwed to the wall. Timber is the usual material for shelving, this can be faced with a large variety of modern plastic finishes. Shelves are classed as solid or slat, the latter being made of 45 x 20 mm slats, spaced about 20 mm apart, and are used mainly in airing and drying cupboards; typical shelf details are shown in Fig. III.26.

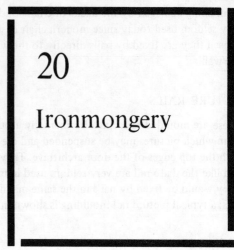

20

Ironmongery

Ironmongery is a general term which is applied to builder's hardware and includes such items as nails, screws, bolts, hinges, locks, catches, window and door fittings.

NAILS

A nail is a fixing device which relies on the grip around its shank and the shear strength of its cross-section to fulfil its function. It is therefore important to select the right type and size of nail for any particular situation. Nails are specified by their type, length and diameter range given in millimetres. The diameter range is comparable to the old British Standard Wire Gauge. The complete range of nails is given in BS 1202. Steel is the main material used in the manufacture of nails, other metals used are copper and aluminium alloy.

Nails in common use are:

Cut clasp nails: made from black rolled steel and used for general carcassing work.

Cut floor brads: made from black rolled steel and used mainly for fixing floor boards, because their rectangular section reduces the tendency of thin boards splitting.

Round plain head: also known as round wire nails and made in a wide variety of lengths, used mainly for general carpentry work but have a tendency to split thin members.

Oval brad head: made with an oval cross section to lessen the risk of splitting the timber members, used for the same purpose as round wire nails but have the advantage of being able to be driven below the surface of the timber. A similar nail with a smaller head is also produced and is called an oval lost head nail.

Clout nails: also called slate nails, they have a large flat head and are used for fixing tiles and slates. The felt nail is similar but has a larger head and is only available in lengths up to 38 mm.

Panel pins: fine wire nails with a small head which can be driven below the surface, used mainly for fixing plywood and hardboard.

Plasterboard nails: the holding power of these nails with their countersunk head and jagged shank makes them suitable for fixing ceiling and similar boards.

WOOD SCREWS

A wood screw is a fixing device used mainly in joinery and relies upon its length and thread for holding power and resistance to direct extraction. For a screw to function properly it must be inserted by rotation and not be driven in with a hammer. It is usually necessary to drill pilot holes for the shank and/or core of the screw.

Wood screws are manufactured from cold drawn wire, steel, brass, stainless steel, aluminium alloy, silica bronze and nickel-copper alloy. In addition to the many different materials a wide range of painted and plated finishes are available such as an enamelled finish known as black japanned. Plated screws are used mainly to match the fitting which they are being used to fix and include such finishes as galvanised steel, copper plated, nickel plated and bronze metal antique.

Screws are specified by their material, type, length and gauge. The screw gauge is the diameter of the shank and is designated by a number; but, unlike the gauge used for nails, the larger the screw gauge number the greater the diameter of the shank. Various head designs are available for all types of wood screws, each having a specific purpose:

Countersunk head: the most common form of screw head and is used for a flush finish, the receiving component being countersunk to receive the screw head.

Raised head: used mainly with good quality fixtures, the rounded portion of the head with its mill slot remains above the surface of the fixture ensuring that the driving tool does not come into contact with the surface causing damage to the finish.

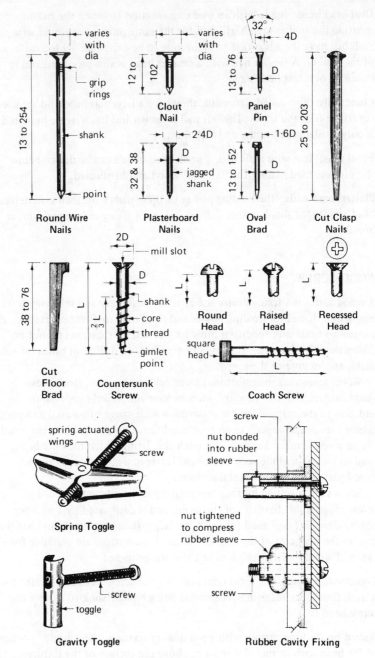

Fig. III.27 Nails, screws and cavity fixings

Round head: the head being fully exposed makes these screws suitable for fittings of material which is too thin to be countersunk.

Recessed head: screws with a countersunk head and a recessed cruciform slot giving a positive drive with a specially shaped screwdriver.

Coach screws: made of mild steel with a square head for driving in with the aid of a spanner and used mainly for heavy carpentry work.

CAVITY FIXINGS

Various fixing devices are available for fixing components to thin materials of low structural strength such as plasterboard and hardboard. Cavity fixings are designed to spread the load over a wide area of the board, typical examples are:

Steel spring toggles: spring actuated wings open out when the toggle fixing has been inserted through a hole in the board and spread out on the reverse side of the board. Spring toggles are specially suited to suspend fixtures from a ceiling.

Steel gravity toggles: when inserted horizontally into a hole in the board the long end of the toggle drops and is pulled against the reverse side of the board when the screw is tightened.

Rubber cavity fixings: a rubber or neoprene sleeve, in which a nut is embedded, is inserted horizontally through a hole in the board, the tightening of the screw causing the sleeve to compress and grip the reverse side of the board. This fixing device forms an airtight, waterproof and vibration resistant fixing.

Typical examples of nails, screws and cavity fixings are shown in Fig. III.27.

HINGES

Hinges are devices used to attach doors, windows and gates to a frame, lining or post so that they are able to pivot about one edge. It is of the utmost importance to specify and use the correct number and type of hinge in any particular situation to ensure correct operation of the door, window or gate. Hinges are classified by their function, length of flap, material used and sometimes by the method of manufacture. Materials used for hinges are steel, brass, cast iron, aluminium and nylon with metal pins. Typical examples of hinges in common use are:

Steel butt hinge: most common type in general use and are made from steel strip which is cut to form the flaps and is pressed around a steel pin.

Steel double flap butt hinge: similar to the steel butt hinge but is made from two steel strips to give extra strength.

Rising butt hinge: used to make the door level rise as it is opened to clear carpets and similar floor coverings. The door will also act as a gravity self closing door when fitted with these butts which are sometimes called skew butt hinges.

Parliament hinge: a form of butt hinge with a projecting knuckle and pin enabling the door to swing through 180° to clear architraves and narrow reveals.

Tee hinge: sometimes called a cross garnet, these hinges are used mainly for hanging matchboarded doors where the weight is distributed over a large area.

Band and hook: a stronger type of tee hinge made from wrought steel and is used for heavy doors and gates. A similar hinge is produced with a pin which projects from the top and bottom of the band and is secured with two retaining cups screwed to the post, these are called reversible hinges.

Typical examples of common hinges are shown in Fig. III.28.

LOCKS AND LATCHES

Any device used to keep a door in the closed position can be classed as a lock or latch. A lock is activated by means of a key whereas a latch is operated by a lever or bar. Latches used on lightweight cupboard doors are usually referred to as catches. Locks can be obtained with a latch bolt so that the door can be kept in the closed position without using a key, these are known as deadlocks.

Locks and latches are either fixed to the face of the door with a staple or keep fixed to the frame when they are termed rim locks or latches. If they are fixed within the body of the door they are called mortice locks or latches. When this form of lock or latch is used the bolts are retained in mortices cut behind the striking plate fixed to the frame (see Fig. III.29).

Cylindrical night latches are fitted to the stile of a door and a connecting bar withdraws the latch when the key is turned. Most night latches have an internal device to stop the bolt being activated from outside by the use of a key.

Door handles, cover plates and axles used to complete lock or latch fittings are collectively called door furniture and are supplied in a wide range of patterns and materials.

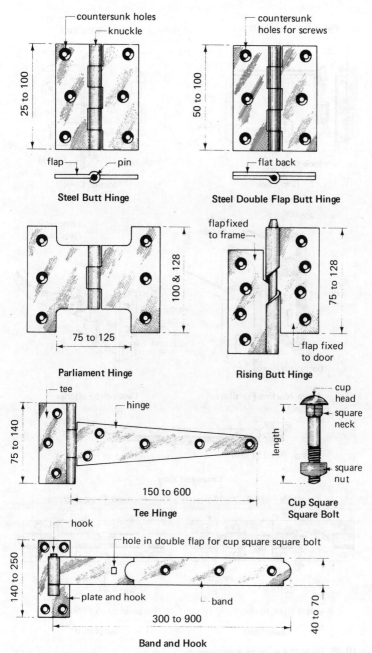

Fig. III.28 Typical hinges

201

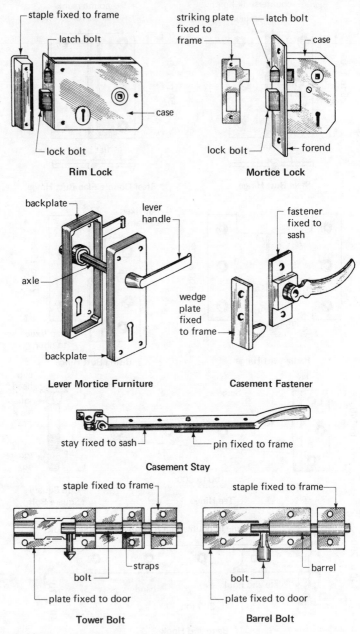

Fig. III.29 Door and window ironmongery

DOOR BOLTS

Door bolts are security devices fixed to the inside faces of doors and consist of a slide or bolt operated by hand to locate in a keep fixed to the frame. Two general patterns are produced: the tower bolt, which is the cheapest form; and the stronger, but dearer, barrel bolt. The bolt of a tower bolt is retained with staples or straps along its length, whereas in a barrel bolt it is completely enclosed along its length (see Fig. III.29).

CASEMENT WINDOW FURNITURE

Two fittings are required for opening sashes, the fastener, which is the security device, and the stay, which holds the sash in the opened position. Fasteners operate by the blade being secured in a mortice cut into the frame or by the blade locating over a projecting wedge or pin fixed to the frame (see Fig. III.29). Casement stays can be obtained to hold the sash open in a number of set positions by using a pin fixed to the frame and having a series of locating holes in the stay or they can be fully adjustable by the stay sliding through a screw down stop fixed to the frame (see Fig. III.29).

LETTER PLATES

These are the hinged covers attached to the outside of a door, which covers the opening made to enable letters to be delivered through the door. The minimum opening size for letter plates recommended by the Post Office is 200 x 45 mm; the bottom of the opening should be sited not lower than 750 mm from the bottom edge of the door and not higher than 1 450 mm to the upper edge of the opening. A wide range of designs are available in steel, aluminium alloy and plastic, some of which have a postal knocker incorporated in the face design.

BS 3287 covers builder's hardware for housing and gives recommendations for materials, finishes and dimensions to a wide range of ironmongery items not covered by a specific standard. These include such fittings as finger plates, cabin hooks, gate latches, cupboard catches and drawer pulls.

21
Painting and decorating

The application of coats of paint to the elements, components, trims and fittings of a building has two functions. The paint will impart colour and, at the same time, provide a protective coating which will increase the durability of the member to which it has been applied. The covering of wall and ceiling surfaces with paper or fabric is basically to provide colour, contrast and atmosphere. To achieve a good durable finish with any form of decoration the preparation of the surface and the correct application of the paint or paper is of the utmost importance.

Paint

Paint is a mixture of a liquid or medium and a colouring or pigment. Mediums used in paint manufacture range from thin liquids to stiff jellies and can be composed of linseed oil, drying oils, synthetic resins and water. The various combinations of these materials forms the type of class of paint. The medium's function is to provide the means of spreading and binding the pigment over the surface to be painted. The pigment provides the body, colour and durability of the paint. White lead is a pigment which gives good durability and moisture resistance but it is poisonous, therefore its use is confined mainly to priming and undercoating paints. Paints containing a lead pigment are required by law to state this fact on the can. The general pigment used for finishing paint is titanium dioxide which gives good obliteration of the undercoating but is not poisonous.

OIL BASED PAINTS

Priming paints: these are first coat paints used to seal the surface, protect the surface against damp air, act as a barrier to prevent any chemical action between the surface and the finishing coats and to give a smooth surface for the subsequent coats. Priming paints are produced for application to wood, metal and plastered surfaces.

Undercoating paints: these are used to build up the protective coating and to provide the correct surface for the finishing coat(s). Undercoat paints contain a greater percentage of pigment than finishing paints and as a result have a matt or flat finish. To obtain a good finishing colour it is essential to use an undercoat of the type and colour recommended by the manufacturer.

Finishing paints: a wide range of colours and finishes including matt, semi-matt, eggshell, satin, gloss and enamel are available. These paints usually contain a synthetic resin which enables them to be easily applied, quick drying and have good adhesive properties. Gloss paints have less pigment than the matt finishes and consequently less obliterating power.

POLYURETHANE PAINTS

These are quick drying paints based on polyurethane resins giving a hard heat resisting surface. They can be used on timber surfaces as a primer and undercoat but metal surfaces will require a base coat of metal primer, the matt finish with its higher pigment content is best for this 'one paint for all coats' treatment. Other finishes available are gloss and eggshell.

WATER BASED PAINTS

Most of the water based paints in general use come under a general classification of emulsion paints: they are quick drying and can be obtained in matt, eggshell, semi-gloss and gloss finishes. The water medium has additives such as polyvinyl acetate and alkyd resin to produce the various finishes. Except for application to iron work, which must be primed with a metal primer, emulsion paints can be used for priming, undercoating and as a finishing application. Their general use is for large flat areas such as ceilings and walls.

VARNISHES AND STAINS

Varnishes form a clear, glossy or matt, tough film over a surface and are a solution of resin and oil, their application being similar to oil based paints. The type of resin used, together with the correct ratio of oil content, forms the various durabilities and finishes available.

Stains can be used to colour or tone the surface of timber before applying a clear coat of varnish; they are basically a dye in a spirit and are therefore quick drying.

Paint supply

Paints are supplied in metal containers of 5 litres, 2·5 litres, 1 litre, 500 millilitres and 250 millilitres capacity and are usually in one of the 88 colours recommended for building purposes in BS 4800. BS 381C gives 99 colours for specific purposes such as standard identification colours used by HM services.

Knotting

Knots or streaks in softwood timber may exude resins which may soften and discolour paint finishes; generally an effective barrier is provided by two coats of knotting applied before the priming coats of paint. Knotting is a uniform dispersion of shellac in a natural or synthetic resin in a suitable solvent. It should be noted that the higher the grade of timber specified, with its lower proportion of knots, the lower will be the risk of resin disfigurement of paint finishes.

Wallpapers

Most wallpapers consist of a base paper of 80% mechanical wood pulp and 20% unbleached sulphite pulp printed on one side with a colour and/or design. Papers with vinyl plastic coatings making them washable are also available, together with a wide range of fabrics suitably coated with vinyl to give various textures and finishes for the decoration of walls and ceilings.

The preparation of the surface to receive wallpaper is very important, ideally a smooth clean and consistent surface is required. Walls which are poor can be lined with a light-weight lining paper for traditional wallpaper and a heavier esparto lining paper for the heavier classes of wallpaper. Lining papers should be applied in the opposite direction to the direction of the final paper covering.

Wallpapers are attached to the surface by means of an adhesive such as starch/flour paste and cellulose pastes which are mixed with water to the manufacturer's recommendations. Heavy washable papers and woven fabrics should always be hung with an adhesive recommended by the paper manufacturer. New plaster and lined surfaces should be treated to provide the necessary 'slip' required for successful paper hanging. A glue size is usually suitable for a starch/flour paste, whereas a diluted cellulose paste should always be used when a cellulose paste is used as the adhesive.

Standard wallpapers are supplied in rolls with a printed width of 533 mm and a length of 10 m giving an approximate coverage of 4·5 m^2—the actual coverage will be governed by the 'drop' of the pattern. Decorative borders and motifs are also available for use in conjunction with standard wallpapers.

Part IV

Water supply
and drainage

22
Domestic water supply

An adequate supply of water is a basic requirement for most buildings for reasons of personal hygiene or for activities such as cooking and manufacturing processes. In most areas a piped supply of water is available from a Public Water Board or Public Utility Company mains supply system.

Water is, in the first instance, produced by condensation in the form of clouds and falls to the ground as rain, snow or hail; it then becomes either surface water in the form of a river, stream or lake, or percolates through the subsoil until it reaches an impervious stratum, or is held in a water bearing subsoil. The water authority by a system of screening, sedimentation, filtration, chlorination, aeration and fluoridation makes the water fit for human consumption before allowing it to enter the mains.

The water company's mains are laid underground at a depth where they will be unaffected by frost or traffic movement. The lay-out of the system is generally a circuit with trunk mains feeding a grid of subsidiary mains for distribution to specific areas or districts. The materials used for main pipes are cast iron and asbestos cement which can be tapped whilst under pressure; a plug cock is inserted into the crown of the mains pipe to provide the means of connecting the communication pipe to supply an individual building.

Terminology
Main: a pipe for the general conveyance of water as distinct from the conveyance to individual premises.

211

Service: a system of pipes and fittings for the supply and distribution of water in any individual premises.

Service pipe: a pipe in a service which is directly subject to pressure from a main, sometimes called the rising main, inside the building.

Communication pipe: that part of the service pipe which is vested in the water undertaking.

Distribution pipe: any pipe in a service conveying water from a storage cistern.

Cistern: a container for water in which the stored water is under atmospheric pressure.

Storage cistern: any cistern other than a flushing cistern.

Tank: a rectangular vessel completely closed and used to store water.

Cylinder: a closed cylindrical tank.

Cold water supply

The water company will provide from their mains tapping plug cock a communication pipe to a stop valve and protection chamber just outside the boundary; this is a chargeable item to the building owner. The goose neck bend is included to relieve any stress likely to be exerted on the mains connection.

A service pipe is taken from this stop valve to an internal stop valve, preferably located just above floor level on an internal wall or at least 600 mm from an external wall. The stop valve should have a drain off valve incorporated in it, or just above it, so that the service pipe or rising main can be drained.

Care must be taken when laying a service pipe that it is not placed in a position where it can be adversely affected by frost, heavy traffic or building loads. A minimum depth of 750 mm is generally recommended for supplies to domestic properties; where the pipe passes under a building it should be housed in a protective duct or pipe suitably insulated within 750 mm of the floor level (see Fig. IV.1).

Suitable materials for service pipes are copper, PVC, polythene, lead and galvanised steel. Copper service pipes can be laid on and covered by a layer of sand to prevent direct contact with the earth or alternatively wrapped with a suitable proprietary insulating material. Plastic coated copper pipes are also available for underground pipework. Steel pipes should have a similar protection but plastic pipes are resistant to both frost and corrosion.

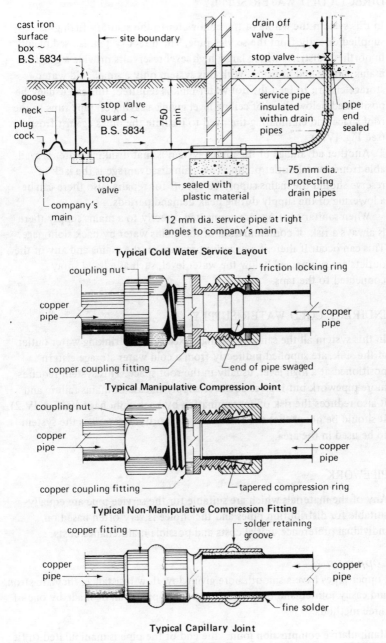

Typical Cold Water Service Layout

Typical Manipulative Compression Joint

Typical Non-Manipulative Compression Fitting

Typical Capillary Joint

Fig. IV.1 Supply pipe and copper pipe joint details

DIRECT COLD WATER SUPPLY

In this system the whole of the cold water to the sanitary fittings is supplied directly from the service pipe. The direct system is used mainly in northern districts where large, high level reservoirs provide a good mains supply and pressure. With this system only a small cold water storage cistern to feed the hot water tank is required; this can usually be positioned below the roof ceiling level giving a saving on pipe runs to the roof space and eliminating the need to insulate the pipes against frost (see Fig. IV.2).

Another advantage of the direct system is that drinking water is available from several outlet points. The main disadvantage is the lack of reserve should the mains supply be cut off for repairs, also there can be a lowering of the supply during peak demand periods.

When sanitary fittings are connected directly to a mains supply there is always a risk of contamination of the mains water by back siphonage. This can occur if there is a negative pressure on the mains and any of the outlets are submerged below the water level, such as a hand spray connected to the taps.

INDIRECT COLD WATER SUPPLY

In this system all the sanitary fittings, except for a drinking water outlet at the sink, are supplied indirectly from a cold water storage cistern positioned at a high level, usually in the roof space. This system requires more pipework but it gives a reserve supply in case of mains failure and it also reduces the risk of contamination by back siphonage (see Fig. IV.2). It should be noted that the local water authority determines the system to be used in the area.

PIPEWORK

Any of the materials which are suitable for the service pipe are equally suitable for distribution pipes and the choice is very often based on individual preference, initial costs and possible maintenance costs.

Copper pipes

Copper pipes have a smooth bore giving low flow resistance, they are strong and easily jointed and bent. Joints in copper pipes can be made by one of three methods:

Manipulative compression joint: the end of the pipe is manipulated to fit into the coupling fitting by means of a special tool. No jointing material is required and the joint offers great resistance to being withdrawn. It is

214

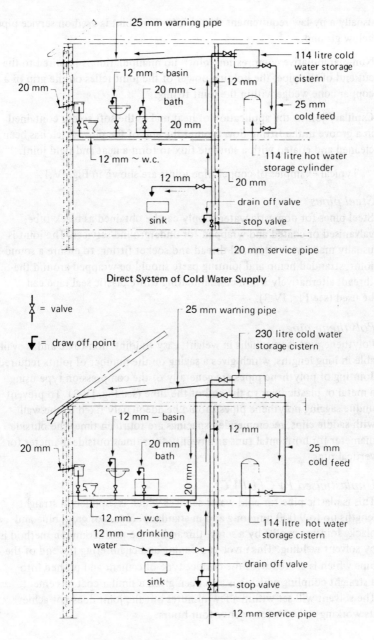

Fig. IV.2 Cold water supply systems

usually a by-law requirement that this type of joint is used on service pipes below ground.

Non-manipulative compression joint: no manipulation is required to the cut end of the pipe, the holding power of the joint relies on the grip of a copper cone wedge within the joint fitting.

Capillary joint: the application of heat makes the soft solder contained in a groove in the fitting flow around the end of the pipe which has been cleaned and coated with a suitable flux to form a neat and rigid joint.

Typical examples of copper pipe joints are shown in Fig. IV.1.

Steel pipes

Steel pipes for domestic water supply can be obtained as black tube, galvanised or coated and wrapped for underground services. The joint is usually made with a tapered thread and socket fitting, to ensure a sound joint, stranded hemp and jointing paste should be wrapped around the thread; alternatively a non-contaminating white plastic seal tape can be used (see Fig. IV.3).

Polythene pipe

Polythene pipe is very light in weight, easy to joint, non-toxic and is available in long lengths, which gives a saving on the number of joints required. Jointing of polythene pipes are generally of the compression type using a metal or plastic liner to the end of the tube (see Fig. IV.3). To prevent undue sagging polythene pipes should be adequately fixed to the wall with saddle clips, recommended spacings are fourteen times the outside diameter for horizontal runs and twenty-four times outside diameter for vertical runs.

Unplasticised PVC (UPVC)

This is plastic pipe for cold water services which is supplied in straight lengths up to 9 000 mm long and in standard colours of grey, blue and black. Jointing can be by a screw thread but the most common method is by solvent welding. This involves cleaning and chamfering the end of the pipe which is coated with the correct type of cement and pushed into a straight coupling which has also been given a similar coat of cement. The solvent will set within a few minutes but the joint does not achieve its working strength for twenty-four hours.

COLD WATER STORAGE CISTERNS

The size of cold water storage cisterns for dwelling houses will depend upon the reserve required and whether the cistern is intended to feed a hot water system. Minimum actual capacities recommended in model water by-laws are 115 litres for cold water storage only and 230 litres for cold and hot water services.

Cisterns should be adequately supported and installed in such a position to give reasonable access for maintenance purposes. The cistern must be installed so that its outlets are above the highest discharge point on the sanitary fittings since the flow is by gravity. If the cistern is housed in the roof the pipes and cistern should be insulated against possible freezing of the water; pre-formed casings of suitable materials are available to suit most standard cistern sizes and shapes. The inlet and outlet connections to the cistern should be on opposite sides to prevent stagnation of water—a typical cistern arrangement together with recommended dimensions for outlets are shown in Fig. IV.3.

BS 417 defines the sizes of galvanised mild steel cisterns which can be used for cold water storage. They have a limited life of 15-20 years due to the breakdown of the protective zinc coating, and their anticipated life can be extended by coating internally with two coats of black bitumen solution.

Plastic cisterns have many advantages over the traditional mild steel cisterns; they are non-corrosive, rot proof, frost resistant and have good resistance to mechanical damage. Materials used are polythene, polypropylene and glass fibre: these cisterns are made with a wall thickness to withstand the water pressure and have an indefinite life. Some forms of polythene cisterns can be distorted to enable them to be passed through an opening of 600 x 600 mm which is a great advantage when planning access to a roof space.

Ball valves

Every pipe supplying a cold water storage cistern must be fitted with a ball valve to prevent an overflow. The ball valve must be fitted at a higher level than the overflow to prevent it becoming submerged and creating the conditions where back siphonage is possible. A ball valve is designed to automatically regulate the supply of water by a floating ball closing the valve when the water reaches a predetermined level.

Two valves are in common use for domestic work, namely the Portsmouth valve and the Garston or BRS valve. The Portsmouth valve has a horizontal piston or plunger which closes over the orifice of a diameter to suit the pressure; high, medium and low pressure valves are available (see Fig. IV.3). The BRS valve is a diaphragm valve which closes over an interchangeable nylon nozzle orifice. This type of valve is quieter in operation, easily adjustable and less susceptible to the corrosion trouble caused by a sticking piston—this is one of the problems that can be encountered with the Portsmouth valve (see Fig. IV.3).

217

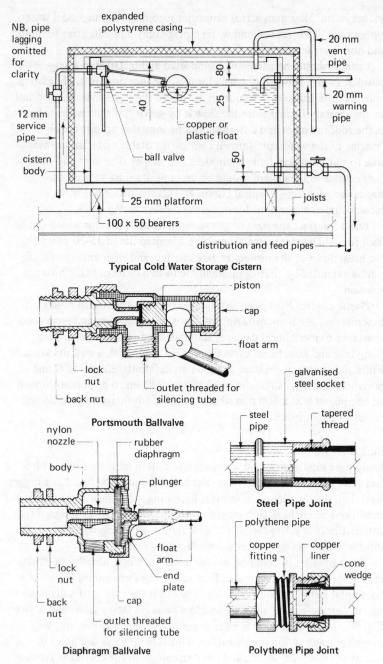

Fig. IV.3 Cisterns, ball-valves and joints

Labels in figure:

NB. pipe lagging omitted for clarity

expanded polystyrene casing

20 mm vent pipe

20 mm warning pipe

12 mm service pipe

copper or plastic float

ball valve

cistern body

25 mm platform

joists

100 × 50 bearers

distribution and feed pipes

Typical Cold Water Storage Cistern

piston

cap

float arm

lock nut

back nut

outlet threaded for silencing tube

Portsmouth Ballvalve

galvanised steel socket

steel pipe

tapered thread

Steel Pipe Joint

nylon nozzle

rubber diaphragm

body

plunger

float arm

end plate

cap

lock nut

back nut

outlet threaded for silencing tube

Diaphragm Ballvalve

polythene pipe

copper fitting

copper liner

cone wedge

Polythene Pipe Joint

Hot water supply

The supply of hot water to domestic sanitary fittings is usually taken from a hot water tank or cylinder. The source of heat is usually in the form of a gas fired, oil fired or solid fuel boiler; alternatives are a back boiler to an open fire or an electric immersion heater fixed into the hot water storage tank. When a quantity of hot water is drawn from the storage tank it is immediately replaced by cold water from the cold water storage cistern. Two main systems are used to heat the water in the tank—these are called the direct and indirect systems. For any hot water system copper or steel pipes are generally used, and care must be taken when connecting copper to steel because of the risk of electrolytic corrosion between dissimilar materials.

DIRECT HOT WATER SYSTEM

This is the simplest and cheapest system; the cold water flows through the water jacket in the boiler where its temperature is raised and convection currents are induced which causes the water to rise and circulate. The hot water leaving the boiler is replaced by colder water descending from the hot water cylinder or tank by gravity thus setting up the circulation. The hot water supply is drawn off from the top of the cylinder by a horizontal pipe at least 450 mm long to prevent 'one pipe' circulation being set up in the vent or expansion pipe. This pipe is run vertically from the hot water distribution pipe to a discharge position over the cold water storage cistern (see Fig. IV.4).

The direct system is not suitable for supplying a central heating circuit or for hard water areas because the pipes and cylinders will become furred with lime deposits. This precipitation of lime occurs when hard water is heated to temperatures of between 50 and 70°C, which is the ideal temperature range for domestic hot water supply.

INDIRECT HOT WATER SYSTEM

This system is designed to overcome the problems of furring which occurs with the direct hot water system. The basic difference is in the cylinder design which now becomes a heat exchanger. The cylinder contains a coil or annulus which is connected to the flow and return pipes from the boiler. A transfer of heat takes place within the cylinder and therefore, after the initial precipitation of lime within the primary circuit and boiler, there is no further furring since fresh cold water is not being constantly introduced into the boiler circuit.

The supply circuit from the cylinder follows the same pattern as the direct hot water system, but a separate feed and expansion system is required for the boiler and primary circuit for initial supply, also for any necessary topping up due to evaporation. The feed cistern is similar to a cold water storage cistern but of a much smaller capacity. The water levels in the two cisterns should be equal so that equal pressures act on the indirect cylinder.

A gravity heating circuit can be taken from the boiler, its distribution being governed by the boiler capacity (see Fig. IV.4). Alternatively a small bore forced system of central heating may be installed.

Hot water cylinders and tanks

Galvanised steel tanks, which are rectangular, can be used for any hot water system where space is restricted and the required storage capacity is less than 155 litres; these storage vessels are usually made to the recommendations of BS 417. Cylinders of galvanised mild steel or copper are produced to the recommendations of BS 1565 and BS 1566 respectively. The standard recommends sizes, capacities and positions for screwed holes for pipe connections.

To overcome the disadvantage of the extra pipework involved when using an indirect cylinder, a single feed indirect or 'Primatic' cylinder can be used. This form of cylinder is entirely self contained and is installed in the same manner as a direct cylinder but functions as an indirect cylinder. It works on the principle of the buoyancy of air which is used to form seals between the primary and secondary water systems. When the system is first filled with water the cylinder commences to fill and fully charges the primary circuit to the boiler with water. When the cylinder water capacity has been reached two air seals will have formed, the first being in the upper chamber of the primatic unit and the second in the air vent pipe. These volumes of air are used to separate the primary and secondary water. When the water is heated in the primary system expansion displaces some of the air in the upper chamber to the lower chamber, this is a reciprocating action; the seals transfer from chamber to chamber as the temperature rises and falls.

Any excess air in the primary system is vented into the secondary system, which will also automatically replenish the primary system should this be necessary. As with indirect systems careful control over the heat output of the boiler is advisable to prevent boiling and consequent furring of the pipework. Typical examples of cylinders are shown in Fig. IV.5.

Faults in hot water systems

Unless a hot water system is correctly designed and installed a number of faults may occur such as air locks and noises. Air locks are pockets of trapped air in the system which will stop or slow down the circulation.

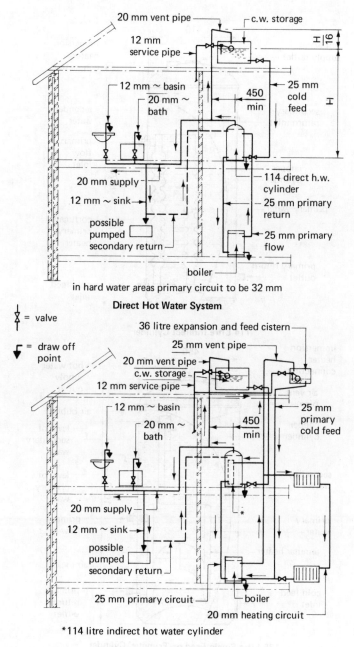

Direct Hot Water System

in hard water areas primary circuit to be 32 mm

Indirect Hot Water System

*114 litre indirect hot water cylinder

Fig. IV.4 Hot water systems

221

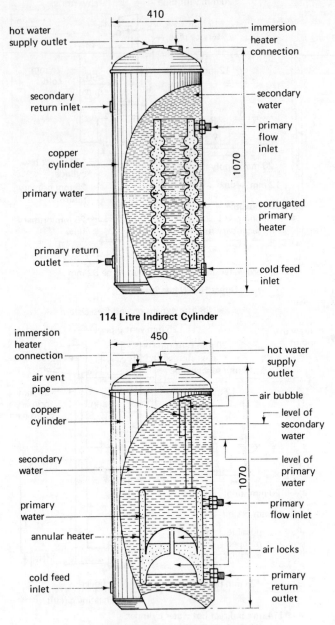

410

hot water supply outlet

immersion heater connection

secondary return inlet

secondary water

primary flow inlet

copper cylinder

1070

primary water

corrugated primary heater

primary return outlet

cold feed inlet

114 Litre Indirect Cylinder

immersion heater connection

450

hot water supply outlet

air vent pipe

air bubble

copper cylinder

level of secondary water

secondary water

1070

level of primary water

primary water

primary flow inlet

annular heater

air locks

cold feed inlet

primary return outlet

135 Litre Single Feed or 'Primatic' Cylinder

Fig. IV.5 Typical hot water cylinders

222

Suspended air in the water will be released when the water is heated, and rise to the highest point. In a good installation the pipes are designed to rise 25 mm in 3 000 mm towards the vent where the air is released through the vent pipe. The most common positions for air locks are sharp bends and the upper rail of a towel rail, the only cure for the latter position is for the towel rail to be vented.

Noises from the hot water system usually indicate a blocked pipe caused by excessive furring or corrosion. The noise is caused by the imprisoned expanded water and the faulty pipe must be descaled or removed, or an explosion may occur.

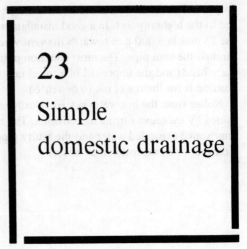

23
Simple domestic drainage

Drainage is a system of pipework, usually installed below ground level, to convey the discharge from sanitary fittings, rainwater gutters and downpipes to a suitable disposal installation. The usual method of disposal is to connect the pipework to the public sewer which will convey the discharges to a Local Authority sewage treatment plant. Alternatives are a small self contained treatment plant on site or a cesspool; the latter is a collection tank to hold the discharge until it can be collected in a special tanker lorry and taken to the Local Authority sewage treatment installation for disposal.

PRINCIPLES OF GOOD DRAINAGE

1. Materials should have adequate strength and durability.
2. Diameter of drain to be as small as practicable: for soil drains the minimum diameter allowed is 100 mm and for surface water the minimum diameter is 75 mm.
3. Every part of a drain should be accessible for the purposes of inspection and cleansing.
4. Drains should be laid in straight runs as far as possible.
5. Drains must be laid to a gradient which will render them efficient. The fall or gradient should be calculated according to the rate of flow, velocity required and the diameter of the drain. Individual domestic buildings have an irregular flow and Maguire's rule will give a gradient with a reasonable velocity. Maguire's rule:

gradient = diameter of pipe (mm)/2·5, therefore for a 100 mm diameter pipe the gradient is 1 in 40 with an approximate velocity of 9·6 l/s. Recommended minimum velocity for small diameter pipes is 4·86 l/s with a maximum velocity of 11·86 l/s. If domestic buildings are linked together with common drains the gradient of 1 in 40 can be considerably reduced by using other formulae for calculating the gradient—but such considerations are beyond the scope of this book.

6. Every drain inlet should be trapped to prevent the entry of foul air into the building, the minimum seal required is 50 mm. The trap seal is provided in many cases by the sanitary fitting itself, rainwater drains need not be trapped unless they connect with a soil drain or sewer.

7. Inspection chambers, manholes, rodding eyes or access fittings should be placed at changes of direction and gradient if these changes would prevent the drain from being readily cleansed.

8. Inspection chambers must also be placed at a junction, unless each run can be cleared from an access point.

9. Junctions between drains must be arranged so that the incoming drain joints at an oblique angle in the direction of the main flow.

10. Avoid drains under buildings if possible; if unavoidable they must be protected to ensure watertightness and to prevent damage. The usual protection methods employed are:
    (i) Encase the drain with 100 mm (minimum) of granular filling.
    (ii) Use cast iron pipes under the building.

11. Drains which are within 1 m of the foundations to the walls of buildings and below the foundation level must be backfilled with concrete up to the level of the underside of the foundations. Drains more than 1 m from the foundations are backfilled with concrete to a depth equal to the distance of the trench from the foundation less 150 mm.

12. Where possible the minimum invert level of a drain should be 450 mm to avoid damage by ground movement and 700 mm for traffic. The invert level is the lowest level of the bore of a drain.

Drainage schemes

The scheme or plan lay-out of drains will depend upon a number of factors:

(a) Number of discharge points.

(*b*) Relative positions of discharge points.

(*c*) Drainage system of the Local Authority sewers.

There are three drainage systems used by Local Authorities in this country and the method employed by any particular authority will determine the basic scheme to be used for the drain runs from individual premises.

COMBINED SYSTEM

All the drains discharge into a common or combined sewer. It is a simple and economic method since there is no duplication of drains. This method has the advantages of easy maintenance, all drains are flushed when it rains and it is impossible to connect to the wrong sewer. The main disadvantage is that all the discharges must pass through the sewage treatment installation, which could be costly and prove to be difficult with periods of heavy rain.

TOTALLY SEPARATE SYSTEM

The most common method employed by Local Authorities; two sewers are used in this method. One sewer receives the surface water discharge and conveys this direct to a suitable outfall such as a river where it is discharged without treatment. The second sewer receives all the soil or foul discharge from baths, basins, sinks, showers and.toilets; this is then conveyed to the sewage treatment installation. More drains are required and it is often necessary to cross drains one over the other. There is a risk of connecting to the wrong sewer and the soil drains are not flushed during heavy rain, but the savings on the treatment of a smaller volume of discharge leads to an overall economy.

PARTIALLY SEPARATE SYSTEM

This is a compromise of the other two systems and is favoured by some Local Authorities because of its flexibility. Two sewers are used, one to carry surface water only and the other to act as a combined sewer. The amount of surface water to be discharged into the combined sewer can be adjusted according to the capacity of the sewage treatment installation.

Soakaways, which are pits below ground level designed to receive surface water and allow it to percolate into the soil, are sometimes used to lessen the load on the surface water sewers. Typical examples of the three drainage systems are shown in Fig. IV.6.

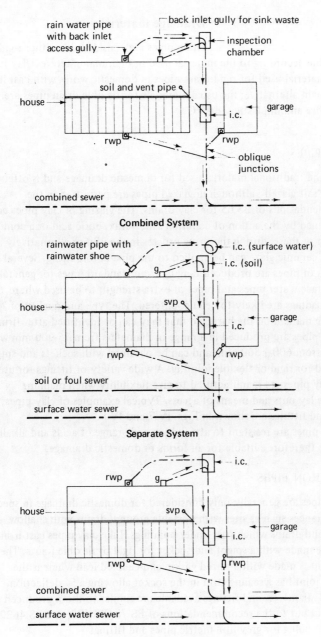

Fig. IV.6 Drainage systems

Drainage materials

Drain pipes are considered as either rigid or flexible according to the material used in their manufacture. Clay is a major material used for rigid drain pipes in domestic work, with cast iron as the main alternative; the usual materials for flexible drain pipes are pitch fibre and unplasticised PVC.

CLAY PIPES

This is the traditional material used for domestic drainage and is often termed 'salt glazed', although unglazed pipes are permitted in the recommendations of BS 65 for clay drains. The glazing of clay pipes can be obtained by the action of common salt, borax, boric acid or a combination of these, during the firing process in the kiln or alternatively a suitable ceramic glaze can be applied to the pipe before firing. Several qualities of pipes are produced ranging from standard pipes for general use, surface water pipes and pipes of extra strength to be used where heavy loadings are likely to be encountered. The type and quality of pipes are marked on the barrel so that they can be identified after firing.

Clay pipes are produced in a range of diameters from 75-900 mm with lengths from 600-1 500 mm, and can be obtained with sockets and spigots prepared for rigid or flexible jointing. A wide variety of fittings for use with clay pipes are manufactured to give flexibility when planning drainage lay-outs and means of access. Typical examples of clay pipes, joints and fittings are shown in Figs. IV.7 and IV.8.

Clay pipes are resistant to attack by a wide range of acids and alkalis; they are therefore suitable for all forms of domestic drainage.

CAST IRON PIPES

These pipes are generally only considered for domestic drainage in special circumstances such as sites with unstable ground, drains with shallow inverts and drains which pass under buildings. Like clay pipes cast iron pipes are made with a spigot and socket for rigid or flexible joints. The rigid joint is made with a tarred gaskin and caulked lead whereas the flexible joint has a sealing strip in the socket allowing a 5° deflection. Lengths, diameters and fittings available are similar to those produced for clay pipes but to the recommendations of BS 78, BS 437, and BS 4622, the latter being for grey iron metric pipes and fittings.

Cast iron pipes are given a protective coating of a hot tar composition or a cold solution of a naphtha and bitumen composition. This coating gives the pipes good protection against corrosion and reasonable durability in average ground conditions.

228

PITCH FIBRE PIPES

Pitch fibre pipes and fittings are made from preformed felted wood cellulose fibres thoroughly impregnated under vacuum and pressure, with at least 65% by weight of coal tar pitch or bituminous compounds. They are suitable for all forms of domestic drainage and because of the smooth bore with its high flow capacity they can generally be laid to lower gradients than most other materials. BS 2760 specifies the strength, composition and resistance requirements for pitch fibre pipes and fittings. Diameters available range from 50-225 mm with general lengths of 2 400 and 3 000mm.

The original joints had a machined $2°$ taper on the ends of the pipe which made a drive fit to machined pitched fibre couplings. These joints are watertight but do not readily accommodate axial or telescopic movement. The snap joint is formed by using a rubber 'D' ring in conjunction with a polypropylene coupling giving a flexible joint and is rapidly superseding the tapered joints (both methods are shown in Fig. IV.7).

UNPLASTICISED PVC PIPES

These pipes and fittings are made from polyvinyl chloride plus additives which are needed to facilitate the manufacture of the polymer and produce a sound, durable pipe. BS 3506 gives the requirements for pipes intended for industrial purposes where BS 4660 covers the pipes and fittings for domestic use. The pipes are obtainable with socket joints for either a solvent welded joint or a ring seal joint (see Fig. IV.7). Like pitch fibre pipes, UPVC pipes have a smooth bore, are light and easy to handle— long lengths reducing to a minimum the number of joints required; they can be jointed and laid in all weathers.

Drain laying

Domestic drains are laid in trenches which are excavated and if necessary timbered in a similar manner to that described in Part I for foundations, the main difference is that drain trenches are excavated to the required fall or gradient. It is good practice to programme the work to enable the activities of excavation, drain laying and backfilling to be carried out in quick succession so that the excavations remain open for the shortest possible time.

The technique used in the laying and bedding of drains will depend upon two factors:

1. Material—rigid or flexible.
2. Joint—rigid or flexible.

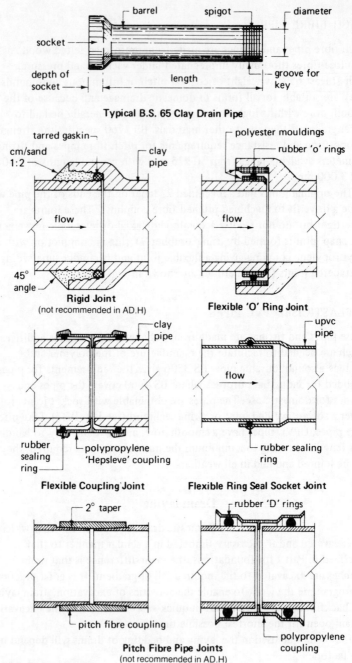

Typical B.S. 65 Clay Drain Pipe

Rigid Joint
(not recommended in AD.H)

Flexible 'O' Ring Joint

Flexible Coupling Joint

Flexible Ring Seal Socket Joint

Pitch Fibre Pipe Joints
(not recommended in AD.H)

Fig. IV.7 Pipes and joints

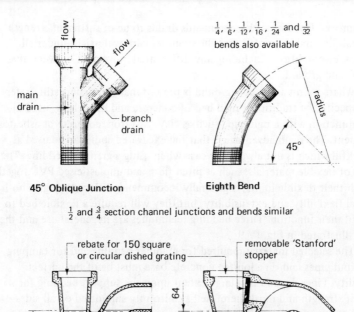

45° Oblique Junction

$\frac{1}{4}$, $\frac{1}{6}$, $\frac{1}{12}$, $\frac{1}{16}$, $\frac{1}{24}$ and $\frac{1}{32}$
bends also available

Eighth Bend

$\frac{1}{2}$ and $\frac{3}{4}$ section channel junctions and bends similar

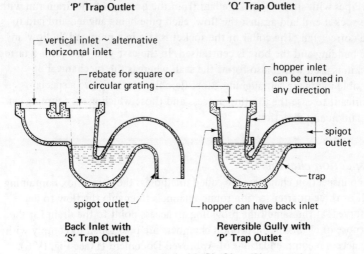

Yard Gully with 'P' Trap Outlet

Access Gully with 'Q' Trap Outlet

Back Inlet with 'S' Trap Outlet

Reversible Gully with 'P' Trap Outlet

all gullies can be supplied with 'P', 'Q', or 'S' trap outlets

Fig. IV.8 Clay drain pipe fittings and gullies

Approved Document H recommends drains to be of sufficient strength, durability and so jointed that the drain remains watertight under all working conditions, including any differential movement between the pipe and ground.

Where a firm and stable ground is present many designers still prefer to specify the traditional rigid bed of concrete and haunching, used in conjunction with a rigid pipe such as clay, cast iron, concrete or asbestos cement. The main advantage is that the excavated spoil can be used as backfill which is not always the case when using a granular bedding. The use of flexible materials, such as pitch fibre and unplasticised PVC together with their flexible joints, is generally recommended since they give both axial flexibility and extensibility, but they will require a flexible bed to fulfil their function. Three bedding techniques are in general use and these are illustrated in Fig. IV.9.

The selected material required for granular bedding and for tamping around pipes laid on a jointed concrete base must be of the correct quality. Pipes depend to a large extent upon the support bedding for their strength and must therefore be uniformly supported on all sides by a material which can be hard compacted. Generally a non-cohesive granular material with a particle size of 5-20 mm is suitable and if not present on site it will have to be 'imported'. Details of the suitability of materials for bedding and surrounding all types of pipes can be found in BS 8301—Code of Practice for Building Drainage.

Pipes with socket joints are laid from the bottom of the drain run with the socket end laid against the flow, each pipe being aligned and laid to the correct fall. The collar of the socket is laid in a prepared 'hollow' in the bedding and the bore is centralised. In the case of a rigid joint a tarred gaskin is used which also forms the seal, whereas the mechanical or flexible joints are self aligning. Most flexible joints require a special lubricant to ease the jointing process and those which use a coupling can be laid in any direction.

Inspection chambers

Deep inspection chambers are called manholes: they are a box containing half or three-quarter section round channels to enable the flow to be observed; at the same time providing an access point to the drain for the purpose of cleansing. Inspection chambers are positioned to comply with the access recommendations of Approved Document H (see Fig. IV.6).

Simple domestic drainage is normally only concerned with shallow inspection chambers up to an invert depth of 1 800 mm. The internal sizing is governed by the depth to invert, the number of branch drains, the diameter of branch drains and the space required for a man to work within

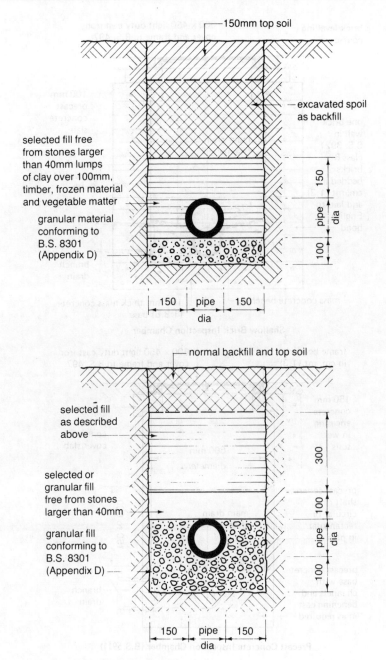

150mm top soil

excavated spoil as backfill

selected fill free from stones larger than 40mm lumps of clay over 100mm, timber, frozen material and vegetable matter

granular material conforming to B.S. 8301 (Appendix D)

150

pipe dia

100

150 | pipe dia | 150

normal backfill and top soil

selected fill as described above

selected or granular fill free from stones larger than 40mm

granular fill conforming to B.S. 8301 (Appendix D)

300

100

pipe dia

100

150 | pipe dia | 150

Fig. IV.9 Typical pipe bedding details

233

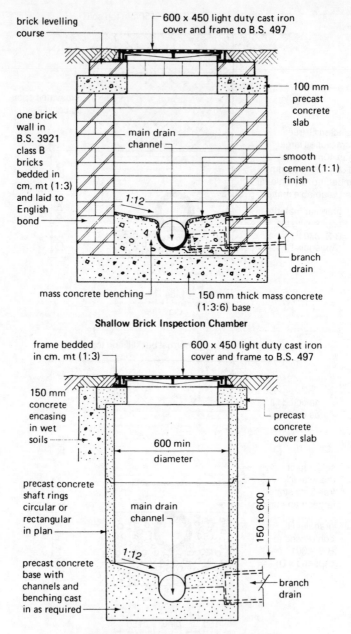

Shallow Brick Inspection Chamber

brick levelling course

600 x 450 light duty cast iron cover and frame to B.S. 497

100 mm precast concrete slab

one brick wall in B.S. 3921 class B bricks bedded in cm. mt (1:3) and laid to English bond

main drain channel

smooth cement (1:1) finish

1:12

branch drain

mass concrete benching

150 mm thick mass concrete (1:3:6) base

frame bedded in cm. mt (1:3)

600 x 450 light duty cast iron cover and frame to B.S. 497

150 mm concrete encasing in wet soils

precast concrete cover slab

600 min diameter

precast concrete shaft rings circular or rectangular in plan

main drain channel

150 to 600

precast concrete base with channels and benching cast in as required

1:12

branch drain

Precast Concrete Inspection Chamber (B.S. 5911)

Fig. IV.10 Typical shallow inspection chambers

234

the inspection chamber. A general guide to the internal sizing of brick built inspection chambers is given in Approved Document H, Table 9 which gives minimum sizes.

Inspection chambers can be constructed of brickwork or of rectangular or circular precast concrete units (see Fig. IV.10). The inspection chamber access covers used in domestic work are generally of cast iron and light duty as defined in BS 497. They have a single seal which should be bedded in grease to form an air-tight joint; double seal covers would be required if an inspection chamber was situated inside the building. Concrete access covers are available for use with surface water inspection chambers.

Ventilation of drains

To prevent foul air from soil and combined drains escaping and causing a nuisance all drains should be vented by a flow of air. A ventilating pipe should be provided at or near the head of each main drain and any branch drain exceeding 10·000 in length. The ventilating pipe can be a separate pipe or the soil discharge stack pipe can be carried upwards to act as a ventilating discharge stack or soil vent pipe. Ventilating pipes should be open to the outside air and carried up for at least 900 mm above the head of any window opening within a horizontal distance of 3·000 from the ventilating pipe which should be finished with a cage or cover which does not restrict the flow of air.

RAINWATER DRAINAGE

A rainwater drainage installation is required to collect the discharge from roofs and paved areas, and convey it to a suitable drainage system. Paved areas, such as garage forecourts or hardstands, are laid to fall so as to direct the rainwater into a yard gully which is connected to the surface water drainage system. A rainwater installation for a roof consists of a collection channel called a 'gutter' which is connected to vertical rainwater pipes. The rainwater pipe is terminated at its lowest point by means of a rainwater shoe for discharge to a surface water drain or a trapped gully if the discharge is to a combined drain (see Fig. IV.11). If a separate system of drainage or soakaways are used it may be possible to connect the rainwater pipe direct to the drains, providing there is an alternative means of access for cleansing.

The materials available for domestic rainwater installations are asbestos, galvanised pressed steel, cast iron and UPVC. The usual materials for domestic work are cast iron and UPVC, the latter being the usual specification for new work.

Cast iron rainwater goods

Cast iron rainwater pipes, gutters and fittings are generally made to the requirements of BS 460 which specifies a half round section gutter with a socket joint in diameters from 75-150 mm and an effective length of 1 800 mm. The gutter socket joint should be lapped in the direction of the flow and sealed with either red lead putty or an approved caulking compound before being bolted together. The gutter is supported at 1 000 to 1 800 mm centres by means of mild steel gutter brackets screwed to the feet of rafters for an open eaves or to the fascia board with a closed eaves.

Cast iron rainwater pipes are also produced to a standard effective length of 1 800 mm with a socket joint which is caulked with red lead putty, run lead or in many cases dry jointed—the pipe diameters range from 50-150 mm. The down pipes are fixed to the wall by means of pipe nails and spacers when the pipes are supplied with ears; or with split ring hinged holderbats when the pipes are supplied without ears cast on. A full range of fittings such as outlets, stopped ends, internal and external angles are available for cast iron half round gutters, and, for the downpipes, fittings such as bends, offsets and rainwater heads are produced.

Unplasticised PVC rainwater goods
The advantages of UPVC rainwater goods over cast iron are:

1. Easier jointing, gutter bolts are not required and the joint is self sealing, generally by means of a butyl or similar strip.
2. Corrosion is eliminated.
3. Decoration is not required, the two standard colours available are black and grey.
4. Breakages are reduced.
5. Better flow properties usually enables smaller sections and lower falls.

Half round gutters are supplied in standard effective lengths of 1 800 and 3 600 mm with diameter range of 100-110 mm; the pipes are supplied in two standard effective lengths of 2 000 and 4 000 mm with diameters of 63, 68 and 75 mm. The gutters, pipes and fittings are generally produced to the requirements of BS 4576. Typical details of domestic rainwater gutter and pipework is shown in Fig. IV.11.

Sizing of pipes and gutters
The sizing of the gutters and downpipes to effectively cater for the discharge from a roof will depend upon:

(a) The area of roof to be drained.
(b) Anticipated intensity of rainfall.
(c) Material of gutter and downpipe.

236

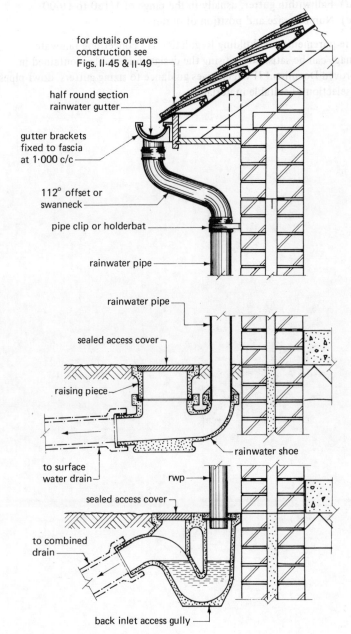

for details of eaves
construction see
Figs. II-45 & II-49

half round section
rainwater gutter

gutter brackets
fixed to fascia
at 1·000 c/c

112° offset or
swanneck

pipe clip or holderbat

rainwater pipe

rainwater pipe

sealed access cover

raising piece

to surface
water drain

rainwater shoe

rwp

sealed access cover

to combined
drain

back inlet access gully

Fig. IV.11 Rainwater pipework and drainage

237

(*d*) Fall within gutter, usually in the range of 1/150 to 1/600.

(*e*) Number, size and position of outlets.

The requirements for Building Regulation H3 concerning rainwater drainage can be satisfied by using the design guide tables contained in Approved Document H which gives guidance to sizing gutters, downpipes and selection of suitable materials.

Bibliography

Relevant BS—British Standards Institution.
Relevant BSCP—British Standards Institution.
Building Regulations 1985—H.M.S.O.
Relevant B.R.S. Digests—H.M.S.O.
Relevant advisory leaflets—D.O.E.
R. Barry. *The Construction of Buildings.* Crosby Lockwood and Sons Ltd.
Mitchells Building Construction Series. B. T. Batsford Ltd.
McKay. *Building Construction*, Vols. 1 to 4. Longman.
Specification. The Architectural Press.
A. J. Elder. *A. J. Guide to the Building Regulations.* The Architectural
   Press.
Relevant A.J. Handbooks. The Architectural Press.
R. Llewelyn Davies and D. J. Petty. *Building Elements.* The Architectural
   Press.
Cecil C. Handisyde. *Building Materials.* The Architectural Press.
L. A. Ragsdale and E. A. Raynham. *Building Materials Technology.*
   Edward Arnold Ltd.
The Green Book on Plastering. British Gypsum Ltd.
The Blue Book on Plasterboard. British Gypsum Ltd.
F. Hall. *Plumbing.* Macmillan.
Leslie Wolley. *Drainage Details.* Northwood Publications.
Relevant manufacturers catalogues contained in the Barbour Index and
   Building Products Index Libraries.

Index

241

242